THE PARSIMONIOUS UNIVERSE

Amor with a Soap Bubble, painting by Rembrandt. (Collections of the Fürst von Liechtenstein, Schloss Vaduz.)

THE PARSIMONIOUS UNIVERSE

SHAPE AND FORM IN THE NATURAL WORLD

STEFAN HILDEBRANDT
ANTHONY TROMBA

COPERNICUS
AN IMPRINT OF SPRINGER-VERLAG

©1996 Springer-Verlag New York, Inc.
Softcover reprint of the hardcover 1st edition 1996
All rights reserved. No part of this publication may be reproduced, stored in a retrieval system, or transmitted, in any form or by any means, electronic, mechanical, photocopying, recording, or otherwise, without the prior written permission of the publisher.

Published in the United States by Copernicus, an imprint of Springer-Verlag New York, Inc.

Copernicus
Springer-Verlag New York, Inc.
175 Fifth Avenue
New York, NY 10010

Library of Congress Cataloging-in-Publication Data

Hildebrandt, Stefan.
 The parsimonious universe : shape and form in the natural world / Stefan Hildebrandt, Anthony Tromba.
 p. cm.
 Includes bibliographical references and index.

 1. Nature (Aesthetics) 2. Form (Philosophy) 3. Motion.
4. Calculus of variations. I. Tromba, Anthony. II. Title.
BH301.N3H555 1995
117–dc20 95-19692

Manufactured in the United States of America.
Printed on acid-free paper.

9 8 7 6 5 4 3 2 1

ISBN 0-387-97991-3 SPIN 10491009

ISBN 978-1-4612-7532-9 ISBN 978-1-4612-2424-2 (eBook)
DOI 10.1007/978-1-4612-2424-2

In memory of Richard Courant

THE BEWITCHMENT

Von Korf finds out his distant cousin is—
 a sorceress
who fashions planets out of herbal fizz;
 and so he hurries, yes,
he hurries there to O-de-lée-de-lizz
 to see the sorceress.

He finds her on a meadow by her home
 and asks her if she be
the one who blows the planets out of foam—
 and if she be the Faërie—
the Faërie from O-de-lée-de-lome?
 Ah, yes, indeed, she be!

She offers him the pitcher and the straw;
 Korf blows,—and from a gleam,
Behold! A wondrous sphere without a flaw
 expands in space supreme,
Expands as if it were a world he saw,
 and not just foam and dream.

CHRISTIAN MORGENSTERN
Translation from the German by Helen and Hans Lewy

Detaching from its stalk, the planet veers
 aloft, and gently up,
and blends into the music of the spheres
 (a Heavenly Choir), floats up . . .
a strain as from the shepherd's pipe appears . . .
 distant tones push up . . .

And in the rounded mirror of this world,
 von Korf perceives with zest,
of all the happy things that ever swirled
 into his mind, the best,–
his mouth agape, beholds his own fair world,
 von Korf, possessed.

He names his cousin "Muse,"–von Korf, possessed,–
 But look! Oh, look again!
For something grabs him by the vest
 and leads him far awain,
Abducts him out of O-de-lá-de-lest
 toward the new domain.

TABLE OF

Preface	x
Prologue: On Form and Shape	1
1 A Grand Scheme of the World	19
2 The Heritage of Ancient Science	41
3 Shortest and Quickest Connections	85
4 A Miracle and Not a Miracle	127

CONTENTS

5 Soap Films: The Amusement of Children and Mathematicians 145

6 Optimal Design 213

Epilogue: Dynamics and Motion 269

References, Comments, and Further Readings 305

Sources of Quotations 318

Sources of Illustrations 319

Index 321

PREFACE

Why does nature produce certain forms, and why does it prefer them to other conceivable forms? What accounts for the similarities and the variety of shapes and forms found in the natural world?

This book explores such questions. It is an account of the calculus of variations, an area of mathematics that deals with optimal forms in geometry and nature—with problems of maxima and minima. We can see examples of nature's remarkable regular forms and patterns starting with the illustrations in the Prologue.

The Greek word *mathema*—which means knowledge, cognition, understanding, perception—suggests that the study of mathematics began about 3000 years ago with asking questions about the world. The historical sections of our account show that a large part of the development of mathematics was the result of a desire to comprehend nature. Mathematics, however, is more than the handmaiden of other sciences. It is, as Carl Friedrich Gauss stated, irrelevant whether one applies mathematical knowledge to number theory or to the movement of a lump of matter such as a planet.

Throughout history, mathematicians have pursued their own ideas, whether relevant to society or not, and have enjoyed the beauty of their discoveries and the challenge of mathematical problems. In 1930, the mathematician C. G. J. Jacobi wrote that the only purpose of science is to honor the human spirit. Considering the two faces of science, there seems to be no other way to accept this statement without reservation but to see, as Kepler did, the ultimate purpose of science as the glorification of God through the recognition of the images "that He has designed for the creation of the world to make it the best and the most beautiful, the most similar to Him." No matter how one may view the aim and value of mathematics, we hope to convince our readers that this science is an integral part of our culture.

We have not used the mathematical formulas and symbols that make modern mathematics so effective and powerful. Indispensable as they are for the working mathematician, extensive training is needed to master mathematical language, a language that may seem strange and alienating to the lay reader. Jonathan Swift ridiculed a society dominated by mathematicians when he described the court of the King of Laputa in *Gulliver's Travels*:

> The knowledge I had in mathematics gave me great assistance in acquiring their phraseology, which depended much upon that science and music; and in the latter I was not unskilled. Their ideas are perpetually conversant in lines and figures. If they would, for example, praise the beauty of a woman, or any other animal, they describe it by rhombs, circles, parallelograms, ellipses, and other geometrical terms, or by words of art drawn from music, needless here to repeat. I observed in the king's kitchen all sorts of mathematical and musical instruments, after the figures of which they cut up the joints that were served to his Majesty's table.

This book is a revised and enlarged version of the original published by W. H. Freeman as *Mathematics and Optimal Form* in 1984. Translations in French, German, Dutch, and Spanish appeared subsequently. In the present version we have included many new pictures, and the figures have been redrawn.

We would like to express our thanks to all who have helped us in writing this book: to Brigitte Hildebrandt, for continual support and discussion; to Peter Renz, who suggested this project and provided judicious suggestions as it developed; to our editor Jerry Lyons, who encouraged us to publish this new edition and with his valuable advice helped us to improve it; to many friends, colleagues, and referees, for reading and criticizing our manuscript or parts of it; to Helen and Hans Lewy, for providing a congenial translation of Morgenstern's poem "The Bewitchment"; to Egbert Brieskorn and Stephen Smale, for photographs of crystals; to Berthold Burkhardt, Klaus Bach,

Frei Otto, and their collaborators and students at the Institute for Lightweight Structures in Stuttgart for generously permitting us to use their photographic archives and for assistance and stimulating discussions; to Emil A. Fellmann, for friendly advice and for providing some historical photographs; to Manfred Kage, for his untiring help and permission to select from his archive of photographs; to Michael Meier, for his photograph of the Devil's Post Pile; to Tom Noddy, for the pictures from his show "Bubble Magic"; to Eric Pitts and Klaus Bach, for preparing photographs of minimal surfaces with free boundaries; to Klaus Steffen and Thomas Deussen, for the photographs of minimal-surface systems in the ten possible networks of great circles on the sphere; to B. Korte for the chip designs; to Alan Schoen, for permitting us to take photographs of his models of minimal surfaces; to T. G. Wang, for the drawing of the three-lobed rotating drop; and to Paul Concus and Robert Finn, for the photograph of a liquid in free fall; to David Hoffman for the computer-generated pictures of Wente's surface; to Michael Nauenberg for interesting conversations on dynamics and for pointing out the existence of the Ole Rømer engraving; to the libraries of the universities in Basel, Bonn, Göttingen, Heidelberg, and Leipzig; to the Bayerische Staatsbibliothek, München, and the Forschungs- und Landesbibliothek Gotha; to the Bancroft Library, Berkeley; to the Lick Observatory, the Hale Observatories, the Jet Propulsion Laboratory in Pasadena, the Bildarchiv Preussischer Kulturbesitz in Berlin, the Biblioteca Apostolica Vaticana, the Liebieghaus in Frankfurt/Main, *Scientific American*, *Life* magazine, the Escher Foundation, the Metropolitan Museum in New York, the National Portrait Gallery in London, the Öffentliche Kunstsammlung Basel, the Herzog Anton Ulrich-Museum Braunschweig, the Museo Archeologico Nazionale di Napoli, the Hamburger Kunsthalle, the Museen der Stadt Köln, and the collections of the Fürst von Liechtenstein, Schloss Vaduz for

permission to use their archives and for providing photographs; and last but not least to the Sonderforschungsbereiche 72 and 256 of Bonn University and to the Max Planck Institut für Mathematik, Bonn, for generous and continuous support.

We cannot give credit to all authors from whose work we have drawn our information. With very few exceptions, we have not mentioned the names of contemporary scientists. However, we would never have been able to write this book without the discussions through the years with our friends and colleagues, and we are grateful for their unselfish support.

<div align="right">

STEFAN HILDEBRANDT
ANTHONY TROMBA

</div>

Ballotini balls.

Prologue

On Form and Shape

Since ancient times, the circle and the sphere have been considered the perfect forms in geometry. For the Greeks, they were the symbols of the ultimate symmetry of the divine. What could be better suited to the immutable and eternal motion of the planets than circular motion? For this reason, the Greeks believed that the planets in the heavens must move on perfect circles, though these might in turn move on other circles. The Greek philosopher Xenophanes (circa 565–470 B.C.) disposed of the multitude of gods of popular belief and replaced them by a unique and supreme God, to whom he attributed the shape of a sphere. Aristotle wrote about the ideas of Xenophanes:

Flagellates are unicellular organisms that move by means of one or more flagella (small whips). Some species of flagellates live as solitary individuals; others may form colonies. The volvox algae produce rather sizable colonies; the spherical surface of each colony may contain as many as 20,000 single cells, each carrying a whip. A volvox colony can organize the movement of its whips to enable it to rotate. Truly splendid spheres are generated by the species *Volvox aureus*, a type of oceanic plankton.

A montage of Jupiter with four of its moons. Io is nearest to Jupiter; next comes Europa, then Ganymede, and Callisto. These satellites were discovered by Galileo. Besides these four, at least ten other, much smaller, moons circle Jupiter; a faint ring of particles also surrounds it.

If God is the best of all, then, he claims, there can be only one. For if there are two or even more, then the one cannot anymore be the very best and strongest of all, because each one of the many gods has the same claim to this rank. In fact, godliness and godly power mean superiority, mean to be the best of all, and not to be surpassed. . . . Hence there can only be God, the same from all sides, and everywhere seeing, hearing, and feeling. Otherwise the various parts would be superior and inferior to each other, and this is impossible. Hence such a universal homogeneity of God implies that he has the shape of a sphere. . . . And since God has his existence in everything, uniquely and eternally, and because he is uniformly the same and round as a ball, thus he is neither limited nor unlimited, neither in rest nor in motion.

The spherical form is as appealing today as it was to the ancient Greeks. We admire the perfect roundness of a floating soap bubble blown by a child. Ballotini balls, which are tiny glass balls, have the same immaculate spherical shape. They have long been used in Venetian glass mosaics. Colonies of marvelous spherical symmetry are generated by green flagellates of the genus *Volvox*. Similarly, the planets of our solar system and their satellites are of nearly spherical shape.

The forces that produce spheres can also produce the shapes seen in the accompanying photographs of diatoms—one-celled algae—which are classic objects of microscopy. They are bounded by a siliceous armor of various delicate geometric shapes.

A great variety of marvelous forms in nature are produced by soap films and bubbles. For reasons we will later see, three soap films will always (if free to do so) form angles of 120° to each other, whether they are spanning wire frames or sitting between glass plates.

The photograph at the lower left on the next page is a view of soap films suspended between two parallel glass plates. Because we are seeing the films as straight lines, it is evident that we are looking at their edges, which means that the films meet each plate at a right angle. Such right angles are formed whenever soap films

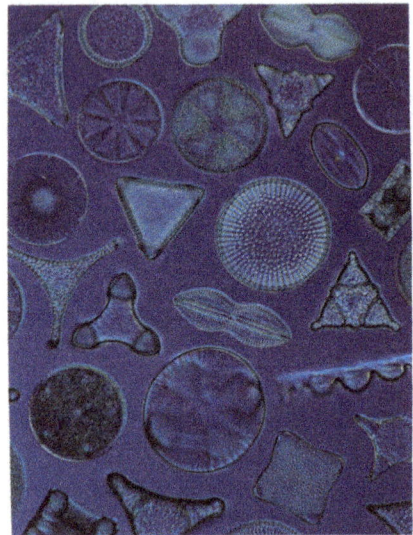
Diatom skeletons, each of which is like a box, with a bottom and a cover.

The species *Arachnoidiscus* USA creates a beautiful disc shell, which might very well serve as the pattern for a gothic window.

The electron microscope reveals the filigreed network of the struts and arches of a diatom, which is penetrated by the living protoplasm.

Soap films between two parallel glass plates. Notice the typical Y-form wherever three films meet. Some cells are bounded by curved films; the bulge is caused by a pressure difference between adjacent cells.

A system of soap films bounded by the twelve edges of an octahedron.

are allowed to arrange their boundaries freely on a supporting surface. For example, the soap films form 90° angles with the sides of a metal strip seen on edge as a thick black line in the lower photograph. Among themselves, however, they meet at 120° angles.

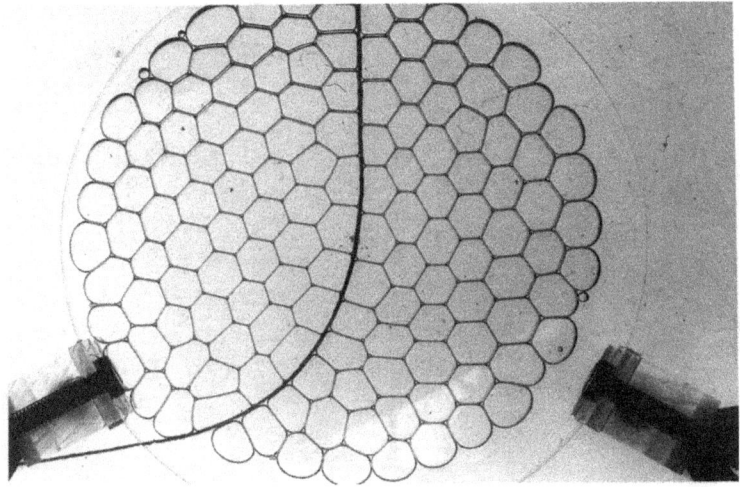

Soap films between two parallel glass plates form 90° angles along a surface (dark line).

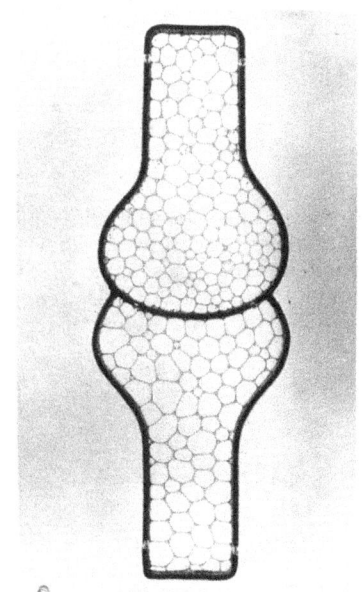

Imitation of bone structures by systems of soap films.

An examination of insect wings and beehive cells reveals their similarity to soap films. It is also fascinating to see the similarities between certain systems of soap films that span polyhedral frames and some unicellular creatures called radiolarians.

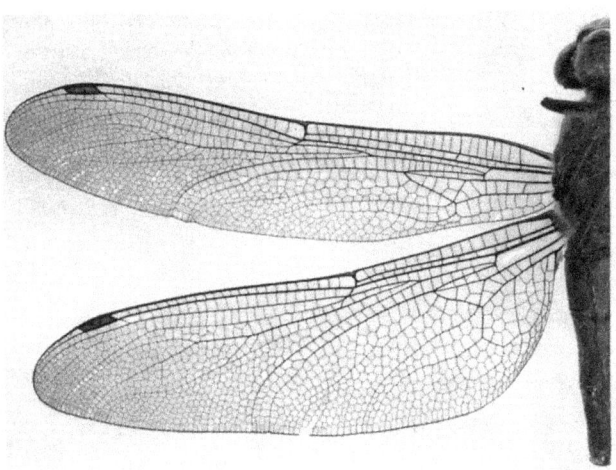

A. Wings of a dragon fly.

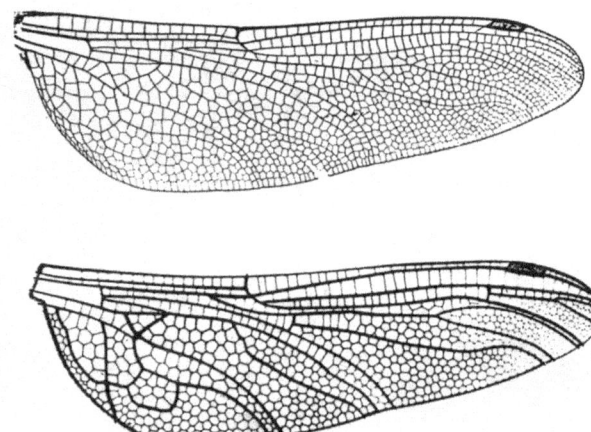

B. A dragon fly's wing, top, can be mimicked by a system of soap films in a suitable frame (thick lines), bottom.

A system of six flat soap films and a soap bubble, suspended within a tetrahedral frame. Such a system resembles the structure of the radiolarian *Callimitra* depicted on the following page.

Drawings by the German biologist Ernst Haeckel of skeletons of the radiolarian named *Callimitra*. Its shape resembles that of a system of six flat soap films holding a bubble, all suspended within a tetrahedral framework (see top of page 8).

The Y-shaped linkage seen in soap film is often found in nature and can cause repeating hexagonal patterns, as in these beehive cells.

A living radiolarian. Haeckel described or depicted 4,314 species and 739 genera of these minute creatures.

The skeleton of a radiolarian.

ON FORM AND SHAPE

Water droplet, with refraction halo.

A phenomenon closely related to soap films and soap bubbles is the formation of a drop, which also kindles our imagination, reminding us of the lovely image of sunlit dew on the petals of a flower.

> So sweet a kiss the golden sun gives not
> To those fresh morning drops upon the rose,

wrote Shakespeare in *Love's Labor's Lost*.

Yet phenomena that stir unpleasant thoughts and feelings, such as cracks, fissures, and crevasses in ice or in drying mud, may also show patterns quite similar to those created by soap films. Whether we see 90° or 120° angles at the points at which the cracks join depends on the material.

Is this an aerial photograph of farmland? No, it is of fissures in a gelatinous preparation of tin oil.

We are also surrounded by continually changing phenomena. Every day we see the Sun rising and setting, just as our ancestors did. On clear nights the Moon and the stars are visible in the sky. The wind may blow, gently rippling the surface of a lake or tumultuously stirring up an ocean's waves. While some of these phenomena follow simple rules, others seem to develop in an intricate or even arbitrary, chaotic way. Who would dare to forecast the weather in the distant future or to predict the path of a whirlwind? Still, even vortices often show certain regularities. The great engineer Theodor von Kármán gave a beautiful mathematical description of the "Kármán vortex street," the formation of a set of whirlpools that can sometimes be seen in rapidly moving liquids or gases. Such a vortex street was depicted by the Japanese artist Hiroshige Utagawa in a woodcut shown below.

A von Kármán vortex street.

Whirlpool at Narúto, by Hiroshige Utagawa.

Smoke ring.

An amazingly structured pattern appears in a smoke ring ejected from a tube. Actually, the result is not a true ring, but a tightly wound, ringlike spiral. Whirl pools of truly cosmic scale are galaxies such as the spiral nebula shown at the right.

Having viewed some of the stunning configurations produced by nature, let us consider this passage from an ancient Buddhist writing:

> . . . a Bodhisattva in accordance with truth knows that form is nothing but holes and cracks, and is indeed like a mass of bubbles, with a nature that has no hardness or solidity.

On Form and Shape 13

Devil's Post Pile. Crack patterns in a vulcanic rock consisting of basalt pillars.

The whirlpool galaxy in Canes Venatici.

A float of bubbles.

Holes and cracks and bubbles: is that all there is to nature's design? Whether or not this is so, in looking at the world around us we often do notice a symmetry of form or a regularity of pattern: there seems to be a certain order in chaos. This order suggests that we ought to look for some simple *universal laws* behind the formation of these configurations. Why does nature prefer this or that form to all other conceivable forms? Why do the celestial bodies look like balls, and not like cubes or pyramids? (Nature does not abhor polyhedral configurations such as cubes and pyramids, as we know from crystalline formations.) Why, we may further ask, are the oil drops floating on our broth circular, and not triangular (like some of the diatoms)?

 Aluminum crystals.

 Palladium crystals.

We do not really hope to find a final answer to such questions: very likely there is none. Most scientists have given up the hope of ever finding a complete answer, an ultimate truth. Instead, they content themselves with unraveling a few principles that will enable them, in a given situation, to predict what forms are preferred by nature over others. This modest approach to understanding complex problems has led to much scientific progress.

Throughout history, people have tried to discover such principles, and it is fascinating to see how our understanding of nature has grown by means of this process. Theory after theory has been devised in order to comprehend nature. Albert Einstein pointed out two reasons why scientific theories are developed:

> New theories are first of all necessary when we encounter new facts which cannot be "explained" by existing theories. But this motivation for setting up new theories is, so to speak, trivial, imposed from without. There is another, more subtle motive of no less importance. This is the striving toward unification and simplification of the premises of the theory as a whole.

The striving for simplicity corresponds to an ideal that most artists, craftsmen, engineers, and scientists have: what you can do, you can do simply. On the surface, this "principle of the economy of means" is merely an aesthetic concept, resting on the conviction that what is done simply is thus done best; yet it also leads to the idea that nature proceeds in the simplest, most efficient way.

Einstein also commented on the human desire to comprehend nature by means of simple, appealing concepts that relate our perception of nature to logical systems of thought:

> There exists a passion for comprehension just as there exists a passion for music. That passion is rather common in children, but gets lost in most people later on. Without this passion, there would be neither mathematics nor natural science. Time and again the passion for understanding has led to the illusion that man is able to comprehend the objective world rationally, by pure thought, without any empirical foundations—in short, by metaphysics. I believe that every true theorist is a kind of tamed metaphysicist, no matter how pure a "positivist" he may fancy himself. The metaphysicist believes that the logically simple is embodied in experienced reality, but that the totality of all sensory experience can be "comprehended" on the basis of a conceptual system built on premises of great simplicity. The skeptic will say that this is a "miracle creed." Admittedly so, but it is a miracle creed which has been borne out to an amazing extent by the development of science.

To no one's surprise, mathematics has played and continues to play a prominent role in the development of such conceptual systems. It appears to be the only tool that allows us to formulate and exploit with precision our insights into the phenomena of the physical world. The British physicist P.A.M. Dirac has expressed this fact in the following way:

> It seems to be one of the fundamental features of nature that fundamental physical laws are described in terms of a mathematical theory of great beauty and power, needing quite a high standard of mathematics for one to understand it. You may wonder: Why is nature constructed along these lines? One can only answer that our present knowledge seems to show that nature is so constructed. We simply have to accept it. One could perhaps describe the situation by saying that God is a mathematician of a very high order, and He used very advanced mathematics in constructing the universe. Our feeble attempts at mathematics enable us to understand a bit of the universe, and as we proceed to develop higher and higher mathematics we can hope to understand the universe better.

In this book we shall consider some of the theories that claim to explain form and motion in our world. We shall see how they came about and what role mathematics has played in their development. In particular, we shall emphasize the *principle of the economy of means*, which throughout has served as a major tool for comprehending physical phenomena. Our goal will be to find an easy-to-understand link between mathematics and physics. The mathematical theory that provides this link is called the *calculus of variations*. Its foundations were laid, and its connection to physics was conceived, during the period of the baroque and the rococo, from the end of the seventeenth century through the eighteenth century—that is, at the very beginning of the development of modern science. We shall therefore start our account by considering some of the mathematical and philosophical ideas of that time.

Pierre-Louis Moreau de Maupertuis (1698–1759), in Northern costume commemorating his Arctic expedition.

1 A Grand Scheme of the World

Throughout history, people have searched for laws to describe the phenomena of our physical world. However, no general principle encompassing all phenomena was proposed until 1744, when the French scientist Pierre-Louis Moreau de Maupertuis put forth his grand scheme of the universe, which became known as the *law of least action*. He published an elaborated version of his ideas in 1746 as "The Laws of Motion and of Rest Deduced from a Metaphysical Principle."

*All that is **superfluous** displeases God and nature. All that displeases God and nature is evil.*
(Dante Alighieri, about 1300)

ACTION AND ECONOMY IN NATURE: MAUPERTUIS'S PRINCIPLE

The "metaphysical principle" of Maupertuis is the assumption that nature always operates with the greatest possible economy. For example, in a homogeneous medium, light would take the shortest possible path. From this idea, he drew the following conclusion, which he stated as his *general principle:*

> If there occurs some change in nature, the amount of action necessary for this change must be as small as possible.

What is this "action" that nature is supposed to consume so thriftily?

Think of a hiker walking along a road, and consider how to describe his action. If he travels 2 kilometers in 1 hour, you would say that he has carried out twice as much action as he would in traveling 2 kilometers in 2 hours. However, you would also say that he carries out twice as much action in traveling 2 kilometers in 2 hours than in traveling 1 kilometer in 1 hour. Altogether then, our hiker, by walking 2 kilometers in 1 hour, carries out *four* times as much action as he would in traveling 1 kilometer in 1 hour.

Using this intuitive idea, we shall define action as the product of distance, velocity, and mass:

$$Action = Mass \times Distance \times Velocity.$$

Mass is included in this definition to account for the hiker's backpack. Moreover, according to Leibniz, the kinetic energy E is given by the formula

$$E = 1/2 \times Mass \times (Velocity)^2;$$

so action has the same physical dimension as

$$Energy \times Time,$$

because velocity is distance divided by time. Thus we have arrived at a quantitative definition of action, which we need to have if we want to give a mathematical formulation of a law of nature that uses the concept of action. It is this formula for action that was used by Maupertuis and that had, incidentally, already been stated by Leibniz. Actually, even our preceding reasoning to motivate this definition of action is taken from one of Leibniz's letters.[1] Very likely Maupertuis had heard about this from a member of the Bernoulli family, a dynasty of Swiss mathematicians who, from 1690 on, made the city of Basel a world center of mathematics. Leibniz was probably also the first to formulate Maupertuis's great metaphysical principle, which we now may paraphrase as,

> Nature always minimizes action.

Maupertuis saw in this principle an expression of the wisdom of the Supreme Being, of God, according to which everything in nature is performed in the most economical way. He wrote,

> What satisfaction for the human spirit that, in contemplating these laws which contain the principle of motion and of rest for all bodies in the universe, he finds the proof of existence of Him who governs the world.

Later we shall explain how the action principle can be used to understand the beautiful forms and patterns we encountered in the Prologue. For the moment, however, we would like to explore the philosophical and scientific ideas that at the time led to the law of least action. How was Maupertuis led to the concept that nature operates with minimal action? Isn't this more a principle of good housekeeping by the Creator than a basis for exact science? It is indeed very interesting to see how this moral principle entered physics. To this end, we shall turn to the late baroque period, when the fundamental mathematical and physical ideas about which we are writing arose. We will see how at that time

[1] To Johann Bernoulli, March 1696.

philosophical, mathematical, and physical concepts developed and merged into a great scheme of the universe that is epitomized in the law of least action. Simultaneously, we will have a glimpse at a fascinating period of history.

PREESTABLISHED HARMONY: THE PHILOSOPHY OF LEIBNIZ

One of the most influential men of the baroque era was Gottfried Wilhelm Leibniz, diplomat, philosopher, mathematician, scientist, and universal scholar. His philosophical ideas were presented in his *Essais de Théodicée sur la Bonté de Dieu, la Liberté de l'Homme et l'Origine du Mal* (Essays on Theodicy: On the Kindness of God, the Freedom of Man, and the Origin of Evil), which appeared in 1710. There he developed the philosophical idea that our world is organized to be *the best of all possible worlds*. This philosophy, well known in educated circles during the first half of the eighteenth century, had many supporters. The most influential of them was probably the German philosopher Christian Wolff (1679–1754). Maupertuis was certainly influenced by Leibniz, although his basic philosophical views differed considerably from those of the great scholar.

Let us look at some of Leibniz's ideas. He had occupied himself in his *Théodicée* with considering how the assumption of an almighty, omniscient, and infinitely good Creator could be compatible with the imperfection of His creation.

By very much simplifying it, we may describe Leibniz's theory as follows. God does not interfere, like a clumsy clockmaker, from time to time with the affairs of this world, to regulate the hands of His clock. On the contrary, God created His world in preestablished harmony. Like a skilled clockmaker looking after his clockwork, God brought the nature of each single part of His world for all eternity into agreement with the nature of all the others; thus all parts are forever in complete harmony with each other. This alone is worthy of God, the most intelligent and almighty being.

Gottfried Wilhelm Leibniz (1646–1716). Painting by Francke, Herzog Anton Ulrich-Museum, Braunschweig.

And God exists, according to Leibniz, for the following reason: God is that being who possesses all properties to the highest degree, insofar as these properties can coexist. This condition is necessary, because some properties, such as holiness and omnipotence, cannot completely coexist. The holiness of God restricts His omnipotence, since it is incompatible with His holiness to do the evil that He could do by means of His omnipotence. Therefore it follows that God may indeed think of all possible worlds, but he may want, and hence may create, only the best among them: the best of all possible worlds.

Thus the existence of this world as the best of all possible worlds is a consequence of the existence of God. Accordingly, any other world would necessarily be less complete than ours; yet the existence of many evils and inequities is not denied: sin and evil do exist. Moreover, Leibniz proves that a world without sin and evil would be impossible.

The best possible world is created in such a way that a preestablished harmony exists between freedom and necessity, between the grace of God and the realm of nature. God as Architect of the world and God as Lord of the universe are in perfect agreement. Leibniz understood perfectly well that this world, being merely the best selection out of what is possible, may be much worse than what we might hope for. However, popular misunderstandings of Leibniz's views reduced his thoughts to this oversimplification: all that exists is good.

If you don't feel that our world is in very good shape, you share the opinion of several contemporaries of Leibniz, and certainly that of Voltaire, the French philosopher, who in 1758 wrote a malicious satire on Leibniz's philosophy titled *Candide, ou l'Optimisme*. It was an immediate sensational success in Europe, and within twenty years forty-two editions appeared. The main characters of the book are the philosopher Pangloss, tutor of the young baroness Cunegonde at the castle of Baron Thunder-ten-tronckh in Westphalia, and his faithful pupil Candide, an illegitimate member

Voltaire (1694–1778)

24 The Parsimonious Universe

of this noble family. These three people experience some rather startling adventures. But let us turn to Voltaire:

> Pangloss taught metaphysico-theologico-cosmolonicology. He proved incontestably that there is no effect without cause, and that in this best of all possible worlds, his lordship's country seat was the most beautiful of mansions and her ladyship the best of all possible ladyships.
>
> "It is proved," he used to say, "that things cannot be other than they are, for since everything was made for a purpose, it follows that everything is made for the best purpose. Observe: our noses were made to carry spectacles; so we have spectacles. Legs were clearly intended for breeches, and so we wear them. Stones were meant for carving and for building houses, and that is why my lord has a most beautiful house; for the greatest baron in Westphalia ought to have the noblest residence. And since pigs were made to be eaten, we eat pork all year around. It follows that those who maintain that all is right talk nonsense; they ought to say, all is for the best."

Investigating the interplay of cause and effect at Castle Thunder-ten-tronckh.

The events that then follow do not strengthen our trust in these assertions of Pangloss, as is to be expected. Candide is kicked out of the castle when Cunegonde wants to experience with him the interplay between cause and effect, and subsequently gets into the Bulgarian army, where he survives the horrors of a battle between the Bulgars and the Abars (who stand for the Prussians and the French).

> Those who have never seen two well-trained armies drawn up for battle can have no idea of the beauty and brilliance of the display. Bugles, fifes, oboes, drums, and salvoes of artillery produced such harmony as Hell itself could not rival. The opening barrage destroyed about six thousand men on each side. Rifle fire that followed rid this best of worlds of about nine or ten thousand villains who infested its surface. Finally, the bayonet provided "sufficient reason" for the death of several thousand more. The total casualties amounted to about thirty thousand. Candide trembled like a philosopher, and hid himself as best he could during this heroic butchery.
>
> When all was over, and the rival kings were celebrating their victory with Te Deums in their respective camps, Candide decided to find somewhere else to pursue his reasoning into cause and effect. He picked his way over piles of the dead and the dying, and reached a neighboring village on the Abar side of the border. It was now no more than a smoking ruin, for the Bulgars had burned it to the ground in accordance with the terms of international law.

Disaster is followed by disaster: Castle Thunder-ten-tronckh is destroyed, its inhabitants murdered or driven away. Cunegonde ends up as a lady of easy virtue in Lisbon, where, having been shipwrecked, Candide and Pangloss arrive just in time to experience the terrible earthquake of October 1, 1755. Thirty thousand men, women, and children are crushed to death under the ruins.

> The following day, while creeping amongst the ruins Candide and Pangloss found something to eat. . . . Some of the citizens whom they had helped gave them as good a dinner as could be managed after such a disaster. The meal was certainly a sad affair, and the guests wept as they ate; but Pangloss consoled them with assurance that things could not be otherwise:

"For all this," said he, "is a manifestation of the rightness of things, since if there is a volcano at Lisbon it could not be anywhere else. For it is impossible for things not to be where they are, because everything is for the best."

In 1762, three years after its first appearance, *Candide* was placed on the papal Index. However, Voltaire had been careful to publish it under a pseudonym, and, as all Europe laughed, he denied vehemently that he was its author.

It is remarkable that, at least initially, the principle of least action shared the fate of its philosophical forebear. The period following its introduction resembled a full-fledged tragicomedy, as once again, to the great amusement of Europe, Voltaire shot off his verbal barbs, this time directed against Maupertuis, for whom the result was disastrous. However, before we turn to this part of our story, we want to describe in very general terms the mathematics on which the principle of least action was to be built.

THE BEGINNING OF THE CALCULUS OF VARIATIONS

In the last third of the seventeenth century, there had been an important development in mathematics. Newton and Leibniz had each independently created *infinitesimal calculus*. This is a mathematical theory that enables one to compute areas of planar figures or volumes of bodies, such as the volume of a wine barrel. It also gives a precise concept of how to define the instantaneous velocity of a moving body and its acceleration. Calculus furnishes computational techniques enabling one to determine the trajectories of bodies from their velocities or accelerations. Computing areas or volumes and determining motions and their velocities or accelerations seem to be very different problems, but they are, surprisingly, quite closely linked. Ancient scientists had already tried to tackle such problems but, except in some special cases, failed. In particular, they had no idea how to define quantitatively the instantaneous velocity of a moving body. Yet this exactly is what is needed if one wants to predict the motions of projectiles or even of the planets. The differential and integral calculus of

Jacob Bernoulli (1654–1705).
Painting by Niklaus Bernoulli, 1694.

Johann Bernoulli (1667–1748).
Engraving by J. J. Haid after a
painting by J. R. Huber, 1740.

Newton and Leibniz provides powerful mathematical methods to attack complex problems involving space and time, making this theory the mathematical foundation of modern science and technology. For instance, no controlled space flights would be possible without the great mathematical achievements of the last centuries that were built upon the foundation of infinitesimal calculus.

However, to be fair to ancient science, we should note that the Greeks invented a rather sophisticated and efficient method, the so-called epicycle theory, to describe the orbits of the Moon or the planets and their motion. This was a purely geometric theory which explained the shape of orbits and the motion of planets *in time* along their orbits. Unfortunately, this theory had to be refined from time to time to account for additional anomalies that were discovered through better observations.

Soon after the discovery of infinitesimal calculus, another new discipline was developed that later became known as "the calculus of variations." The founding fathers of this new field were three world-famous Swiss scientists, all from the city of Basel: the brothers Johann and Jacob Bernoulli; and Johann's student, Leonhard Euler,

who would become the most celebrated mathematician of the eighteenth century. The goals of this new mathematical discipline are closely related to the philosophical ideas just described. They are to find mathematical descriptions of optimal objects and to develop the mathematical techniques needed to determine them.

Like anyone, mathematicians are quite interested in such problems. In daily life, we must regularly decide such questions as which situation is "best" or "worst"; which object has some property to a "highest" or "lowest" degree; what is the "optimal" strategy to maximize pleasure, success, or profit, or to minimize discomfort, failure, or losses. What shape of a car will, for a prescribed interior volume, minimize its resistance to air? What shape of a hull will produce the fastest ship? What form of a house has the least possible surface area for a given interior volume, to ensure that the least possible heat is

View of the city of Basel in the time of the Bernoullis, with the Rhine bridge in the foreground. Etching by W. Herrliberger (1761) after a drawing by E. Büchel.

lost to the outside? (The answer to this last question can help reduce your heating bill. The Eskimos found it a long time ago: the optimal house form is the igloo—that is, the hemisphere.)

How to find summits using a level.

The calculus of variations was developed by the Bernoullis, by Euler, and by the great French mathematician J. L. Lagrange to attack such questions in a systematic way. The basic idea of this mathematical theory is easy to grasp, although the application to specific problems usually is quite involved.

Suppose that you are in a mountainous area in the dark. You want to locate the summits but are equipped with only a small flashlight and a level. How would you proceed? First you would recall that at a summit the tangent plane to the mountain range is horizontal (see accompanying figure), and the same is true for the lowest point of a valley or a saddle point. You could then walk around and use your level to try to locate all the places where the ground is horizontal—that is, has a horizontal tangent plane. If the mountain range is geologically very old, so that the surface of the earth is smooth, and not all rock and full of peaks, you can, in principle, find all the summits in this way. However, you may also find other points with horizontal tangent planes—for example, the mountain passes or mountain pits, all of which you might pass on your way to a summit. *In practice*, no one would really proceed in this manner; but these considerations show how you could solve this problem *in principle*, by employing the idea of the horizontal tangent. This is how the mathematician locates maxima and minima, by first reducing questions of best or worst, *maxima* or *minima*, to a geometric question of finding summits or pits in some mathematically constructed mountain range. However, these are usually not mountains in a three-dimensional world, but in a "higher-dimensional" space. This sounds quite mysterious, as nobody can imagine a seven-dimensional space, not to mention an infinite-dimensional one, but such artificial spaces are technical constructs forced upon mathematicians if they want to approach seemingly diverse problems with a uniform methodology, and they have learned to deal with such spaces technically, even though they can usually visualize mountains only in three-dimensional space.

Joseph Louis Lagrange (1736–1813).

Frontispiece of Voltaire's book on Newtonian physics (1738).

Frontispiece of Lagrange's *Méchanique Analitique* (1788).

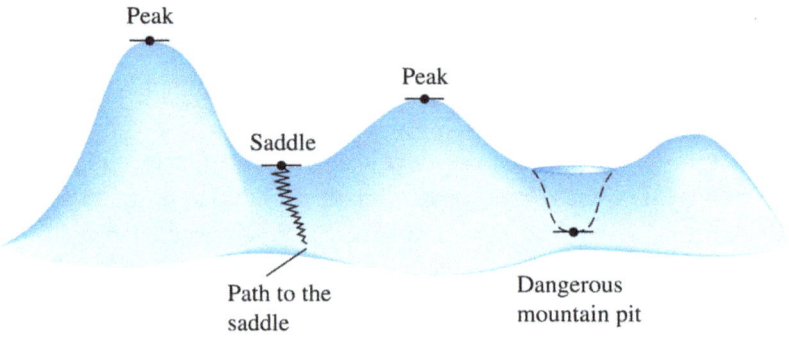

Peaks, pits, and passes.

The mathematician then uses a strategy (such as the one described) to focus attention on a small number of points suspected to be maxima (peaks) or minima (pits). This procedure is similar to that of a detective, who uses all available circumstantial evidence to reduce the number of suspects who might have committed a crime.

From the "principle of the horizontal tangent plane" the mathematician establishes a system of differential equations (called the Euler–Lagrange equations) that must be satisfied by the optimal objects that are being sought, and then attempts to solve them. Finding the solutions of these equations is usually difficult and needs a lot of mathematical knowledge; therefore we will not go into the technical details.

THE MECHANICS OF NEWTON

So far, we have described only the philosophical and mathematical ideas from which the principle of least action originated. However, in the form proposed by Maupertuis, the principle was too vaguely formulated to be applied to difficult problems. Euler and Lagrange elaborated and perfected it into a powerful tool that was to become indispensable for physics. An elegant general version appeared in Lagrange's much admired masterpiece, *Méchnique analitique* (Analytical Mechanics), which was published in 1788, just one year before the outbreak of the French Revolution.

All these developments, however, were based on the revolution in physics that took place when Newton's epoch-making *Philosophiae Naturalis Principia Mathematica* (Mathematical Principles of the Philosophy of Nature) appeared in 1687. This treatise, one of the most influential creations of the human mind, and the foundation of modern physics, is also the solid pillar on which the least-action principle safely rests.

However, Newton's revolutionary ideas were not readily accepted on the continent of Europe, particularly in France, the cultural center of Europe at that time. During the seventeenth century, and until about 1730, continental Europe was dominated by the physical theories of René Descartes. Since the triumph of Newtonian physics, however, his theories have been almost completely forgotten. The Cartesians, as the followers of Descartes were called, vigorously battled Newtonianism, until the French Academy of Sciences agreed to decide the controversy on the basis of facts rather than by debate. It sent out two expeditions, one to the Arctic Circle in Sweden and one to the Equator, to take geodesic measurements that would decide whether the Earth was flattened at the poles, as Newton had predicted from his theory of gravitation, or at the Equator, as the Cartesians had asserted.

The expedition to the north was carried out in 1736–37 by Maupertuis, who had been an early supporter of Newton's ideas. From the geodesic measurements taken in Lapland, it could be

René Descartes (1596–1650). Engraving after a painting by Frans Hals, 1648.

An engraving from Maupertuis's report of his Lapland expedition, published under the title *La Figure de la Terre* in 1738.

Exaggerated drawings of (A) Newton's view of the Earth; (B) Descartes's view of the Earth.

Frontispiece of the French translation of Newton's *Principia* by Gabrielle-Emilie le Tonnelier de Breteuil, Marquise du Châtelet. First edition, second issue, Paris, 1759. (The first issue, dated 1756, was not officially published and is very rare.)

proved that the Earth was flattened at the poles and that Newton was right. This result was confirmed in 1745 by another expedition led by La Condamine.

The great success of his expedition brought Maupertuis instant renown—a fame that was embellished by the two young Lapp girls whom he had brought back to Paris. Evidently they made quite an impression on the Parisians.

Voltaire, with his prodigious pen and quick mind, helped to publicize Newton's great triumph. He congratulated Maupertuis for successfully and simultaneously flattening the globe and the Cassinis (well-known astronomers in Paris and vigorous defenders of the Cartesian view of the Earth's shape). Look once more at Maupertuis's gesture in the figure on page 18, to see how proudly he indicates his success.

Thus the way was cleared for Newtonian physics and for its elegant formulation in terms of variational (optimum) principles. The era of "analytical mechanics" could begin, as the new science of mechanics was later called, with reference to Lagrange's masterpiece. This is what eventually happened.

THE BERLIN CONTROVERSY OVER MAUPERTUIS'S PRINCIPLE

In 1740 Frederic II became King of Prussia, the northeastern part of Germany with Berlin as its capital. He wanted to revive the Academy of Sciences in Berlin and he invited Voltaire to accept its presidency. Voltaire was pleased but declined to come because he preferred to stay at the domicile of his ladylove, the Marquise du Châtelet.[2] Frederic therefore offered the presidency of the Prussian academy to Maupertuis, then famous as the organizer of the expedition to the Arctic Circle. He accepted, but did not assume his position until 1746.

Meanwhile, Euler had come to Berlin in 1741 from St. Petersburg in Russia to become the director of the mathematical-physical section of the Academy. Leonhard Euler (1707–1783), son of a parson, entered the university of his hometown Basel at the early age of thirteen. This was not as uncommon in those days as it may seem. School education in those times was usually restricted to the basics, namely to religion, and—most importantly—to Latin. Moreover, the instruction at universities was, more often than not, at a rather modest level. Leibniz, for example, began his studies at the University of Leipzig, his hometown, when he was fourteen, and he bitterly criticized the shamefully low level of the lectures that he heard. In fact, he later said that he could have contributed to science much earlier if he, like Pascal, had spent his youth in Paris. The young Euler was much luckier, since Johann Bernoulli had early on discovered the talent of the boy and decided to give him private lessons each Saturday. Since at this time Johann was the leading mathematician in Europe, his pupil Euler was probably already quite learned when, at

The beginning of Voltaire's poem for the Marquise du Châtelet.

[2]This remarkable woman had been tutored by some of the outstanding scientists of her time, including J. S. König and A. C. Clairaut. She prepared, with the assistance of Clairaut, the first and only French translation of Newton's *Principia*, to which Voltaire added upon her death in 1749 a historical preface and a poem, "Sur la Physique de Newton."

the age of twenty, he applied for a professorship in physics at the University of Basel. However, Euler was not chosen, and so he accepted a research position at the newly founded Academy of St. Petersburg, where a whole colony of Swiss scientists had already assembled, among them two sons of Johann Bernoulli. There he initiated scientific investigations that did not end until his death in 1783. Remarkably, even after the loss of his eyesight in 1771, he published several books and more than 300 papers. During his entire lifetime he wrote about 900 papers and 20 books, making fundamental contributions to all fields of mathematics as well as to mechanics, optics, astronomy, and even to musicology. Presently, more than seventy volumes of his *opera omnia* have appeared, but the editorial work is still not finished.

By the time he left for Berlin in 1741, Euler had written more than one hundred papers in addition to his *Mechanica* (which appeared in 1736). This was the first presentation of Newtonian mechanics using analytical methods and infinitesimal calculus (Newton and other predecessors of Leibniz had avoided calculus and had written in a more geometric and synthetic fashion). In this work Euler revolutionized the field of mechanics and, together with his second treatise on mechanics, the *Theoria Motus* of 1765, he prepared the ground for Lagrange's great contributions.

By 1741, the political situation in Russia had become quite unpleasant, and Euler therefore accepted King Frederic's invitation to come to Berlin. Although he stayed in Berlin for about 25 years, Euler's relations with the King were never quite satisfactory, and they eventually became unbearable. Therefore, at the invitation of Tsarina Catherine II, he returned to St. Petersburg in 1766, becoming once again a member of the Imperial Russian Academy. Euler is considered to be the father of Russian mathematics.

By 1744, Euler had already rigorously proved that the least-action principle could be used to describe the motion of a point-mass in a conservative field of forces, such as the motion of a planet around the Sun, and he had expressed his conviction that, behind every

phenomenon in our universe, we can find a maximum or minimum rule. This result appeared in an appendix to his *Methodus Inveniendi Lineas Curvas Maximi Minimive Proprietate Gaudentes* (A Method to Find Curved Lines That Enjoy a Maximum or Minimum Property), the first textbook on the calculus of variations and one of the most celebrated books in the history of mathematics. It is certain that Euler had known nothing of Maupertuis's principle when he wrote the appendix to his book in 1743. Moreover, the problems Euler treated were much more difficult than those that Maupertuis would handle in his 1746 paper.

When, in 1746, Maupertuis published his paper on the least-action principle, he was well aware of Euler's achievement, since he briefly described it in the preface to his publication. Then, however, he added the comment, "This remark . . . is a beautiful application of my principle to the movement of the planets," and thus claimed the right of priority for himself.

Euler reacted by attributing the whole priority to Maupertuis, arguing that his own principle of economy was only an "*a posteriori* insight," whereas Maupertuis's law was an "*a priori* conception of the world." Euler has been criticized strongly for this act. Some historians have attributed it to generosity and loyalty to the president of his academy; others have seen it as a pitiful example of moral weakness. Still other commentators feel that either judgment is too simpleminded and that the real reason for Euler's behavior lies in his philosophical attitude and religious conviction. It is not a question on which certainty is possible.

However, Maupertuis's reasoning had several flaws. First, he supported his principle by only a few examples; second, and worse, the principle was not always correct.

Let us take one of Maupertuis's examples, the reflection of light. Here the law of least action turns out to be equivalent to the rule that light propagates along rays in such a way that the time needed by the light to get from the source to the observer is less than that

Frontispiece of Euler's *Methodus Inveniendi* (1744).

which would be required along all other possible paths. (This latter rule for light propagation had already been formulated by Fermat in 1662.) If light travels in a homogeneous isotropic medium, like water or air of the same density, this minimum principle amounts to the still simpler rule that the light path is the shortest possible connection of source and observer, a rule already known to the ancient Greeks.

Consider a spherical mirror in such a perfect medium and a ray starting at the center, M, of the mirror, as shown below. The ray, according to Maupertuis's principle, is reflected and travels to some point P beyond M. It is easy to see that any other path that consists of two straight segments leading from M to the mirror and then to P is shorter than the light ray. If, on the other hand, we reflect light from a convex mirror, the path of light is indeed minimal. Thus whether nature acts thriftily or wastefully depends only on the shape of the mirror.

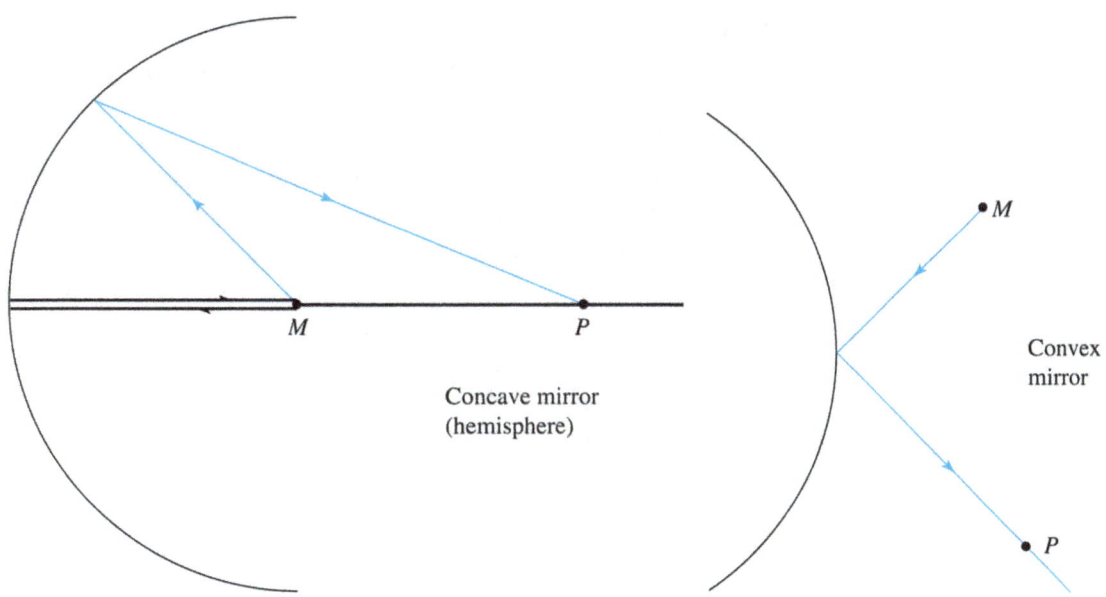

Left: Light reflected from a concave surface. The blue lines show a possible connection of M with P by way of the mirror that is shorter than the actual light ray (as indicated by the heavy black lines). Right: Light reflected from a convex surface.

Frontispiece of the *Maupertuisiana*, the collected diatribes for and against the president of the Berlin Academy.

Frontispiece of the *Docteur Akakia*, together with the scribblings of an unknown hand.

With arguments of this kind, the Chevalier d'Arcy showed in 1749 and in 1752 that Maupertuis's principle was not clearly formulated and, moreover, led to incorrect assertions. The theologic-philosophical construction built on the principle of greatest economy seemed irreparably damaged.

In Berlin, Maupertuis was criticized by a new member of the Academy, the Swiss mathematician Johann Samuel König (1712–1757), who arrived in Berlin in 1750 and claimed that Leibniz had already stated the least-action principle in 1707. A heated controversy arose over whether König's documents to support this claim were forged, as he had only submitted a copy of Leibniz's letter where the action principle had been stated. The Academy asked König to present the original, which he allegedly had seen in the hands of Samuel Henzi, a young Swiss poet. Henzi lived in the city of Bern or, rather, had lived in that city, since in the interim the city fathers of Bern had had him beheaded for his involvement in the revolutionary activities of 1749. Subsequently the assets of Henzi had been seized and then sold or destroyed. Thus Leibniz's letter could not be found, even though King Frederic took it upon himself to address an official inquiry to the city counsel of Bern. Therefore, on April 13, 1752, the Academy declared König's copy of the crucial letter by Leibniz to be a falsification. König reacted by resigning his membership in the Academy, and a heated quarrel started between König, on the one hand, and Maupertuis and Euler, on the other. Voltaire then entered the fray, and in 1752 he attacked Maupertuis in his *Diatribe of Doctor Akakia*. As a result, the literati and encyclopedists of Paris took sides with Voltaire against Maupertuis, and Maupertuis was finished. A sick man, he left King Frederic's court in 1753 to recover his health. Only occasionally did he return to Berlin. He died in 1759 in Basel, at the house of his friend Johann Bernoulli II.

Today the dispute about who first discovered the least-action principle is meaningless, as are most quarrels of this kind. We can give both Euler and Maupertuis credit for its discovery. The true priority, however, very likely belongs to Leibniz because we now know that

he formulated and substantiated the action principle in several letters. Admittedly only in the lost letter did he really state precisely, according to his general teleological principle, that the action should have an extremum (i.e., a maximum or minimum). However, the authenticity of this controversial letter became much more probable when in 1913, among a convolute of other letters by Leibniz, another, hitherto unknown copy was discovered in the famous Gotha library, copied by the same hand and known to be authentic. Yet for posterity Maupertuis remains the man who first proposed the action principle *as a universal law of physics*, from which all other laws should follow, whereas Euler created the mathematical structure that served as a model for all later variational principles.

Recapitulating, we can say that Maupertuis's grand scheme of the world, expressed in the form of the action principle, had its roots in the philosophy of Leibniz. Leibniz had already stated this principle in an even better and more general form, as he did not require action to be only minimized, but possibly maximized. For Leibniz maximum and minimum were of equal value, whereas Maupertuis's metaphysical principle of economy demanded a minimum as its essential ingredient. Yet the action principle could never have advanced to a universal law capable of quantitative applications without the considerable progress eventually made in mathematics and physics. Infinitesimal calculus and Newtonian mechanics were the firm ground on which Euler and Lagrange could establish the action principle.

A copy of Leibniz's letter from 1707 kept in the Gotha library at Friedenstein castle, where it was discovered by Kabitz in 1913. Here, from the last two pages, is Leibniz's formulation of the action principle.

However, the idea that a fundamental order and harmony of the universe existed, an order that could be described by means of mathematics, goes back to the ancient Greeks. This was well known to Leibniz, who had mused on such ideas in Plato's *Phaidon*, and also to many other scientists of the Renaissance and the baroque period. Therefore, our account would be deficient without a look at the heritage of ancient science.

Plato's Academy (mosaic found in Pompeii, Villa of T. Siminius Stephanus, 86 x 85 cm, Naples, Archaeological Museum). With certainty the seven men have been identified as Plato (third from the left) and six other philosophers, who are talking about the universe, the celestial spheres, and the stars. The mosaic shows Plato's Academy, with the city of Athens in the background. It is probably a copy (from the first century B.C.) of a Hellenistic painting which possibly dates from the third century B.C.

2 The Heritage of Ancient Science

Many fundamental ideas in science were conceived in antiquity, and our present way of thinking owes, without doubt, a great deal to our predecessors. One essential idea that modern science has inherited from the classical world is the concept of a fundamental order and harmony to the universe, a harmony that could be reflected in the beauty of mathematical structures. Because Greek mathematics was mainly restricted to geometry, the ancient scientists used geometric models to describe nature.

Thou, Nature, art my goddess; To thy laws my services are bound.
(Shakespeare, *King Lear*, Act I, Scene II, as quoted by Gauss)

Interestingly, optimum principles appeared in ancient science, but only marginally, although it should have been very pleasing for the ancient scholars to think that the harmony of things corresponded to an optimal order in the world. Let us take a closer look at ancient Greece and even earlier civilizations, to begin to see how the geometry of ideal forms was invented.

EGYPTIAN AND BABYLONIAN MATHEMATICS

It is usually said that mathematics developed in Greece in the seventh and sixth centuries B.C., after the Greeks had devised a more or less uniform alphabet (the classical Greek alphabet was commonly accepted only during the fourth century B.C.). However, modern historians admit that our knowledge about the science of this time is by no means well founded. Primary sources do not exist, and all events were recorded only long after they had actually occurred. Even the scientists of antiquity had no sure knowledge of the origin of mathematics and of the first mathematicians. The traditional stories about the two towering figures of early mathematics, Thales of Miletus (about 624–548 B.C.) and Pythagoras of Samos (about 580–500 B.C.) are probably more or less legendary. It now appears certain that the mathematical knowledge commonly ascribed to the early Greek philosophers was, in fact, known to the Egyptians and the Babylonians many centuries before the rise of Greek civilization. One of the main sources of our information on Egyptian mathematics is the famous Papyrus Rhind (seventeenth century B.C.), kept at the British Museum in London, a kind of mathematical handbook. From this and other papyri we have learned that the ancient Egyptians had a kind of decimal system and could very well calculate with fractions. They were also able to solve first- and second-order equations in one variable; they could compute areas of certain geometric figures, such as squares, rectangles, triangles, or trapezia; and they had a reasonably good approximative formula for the area of a circular disk using the value 3.16 for the area of the unit disk, that is, for the number π. They

Engraving in Euclid's *Opera Omnia,* edited by David Gregory, Oxford, 1703, depicting an episode told by the Roman architect Vitruv. The Greek philosopher Aristippos, shipwrecked, discovering geometric figures drawn on the sand, encourages his companions by saying: "Friends, we may well hope, for I see human traces."

Pythagoras: a twelfth-century sculpture on the Cathedral of Chartres.

knew how to compute certain volumes, like those of cubes, rectangular solids, solid cones, cylinders, and pyramids. The ancient Egyptians were even acquainted with the Pythagorean theorem, at least in certain cases, and used it to construct right angles by Pythagorean triples such as 3, 4, 5 ($3^2 + 4^2 = 5^2$) via knotty ropes.

The Sumerian-Babylonian mathematics had reached the highest level among all the cultures preceding the ancient Greeks. It even reached to India and, as some researchers believe, to China. Our knowledge comes from numerous cuneiform scripts on baked clay tiles that have been excavated in Mesopotamia, the region between the Euphrates and Tigris Rivers. In the third millenium B.C., the Sumerians already had a number system using the base 60, which, during the third dynasty of Ur (2400–2000 B.C.), was developed to a sexagesimal position system. It appears that this sexagesimal system was essentially used in mathematical texts, whereas in daily life a

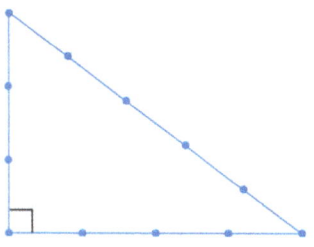

The carpenter's construction of a rectanglar triangle by means of a knotted rope.

decimal system was applied. The Mesopotamian mathematics reached its zenith in the elder Babylonian time (about 2000–1600 B.C.). From this time on Babylonians possessed an excellent knowledge of arithmetic; they could take roots, they had exponential tables, and even the beginnings of logarithmic computation can be seen. The older Babylonian mathematics was basically of an algebraic nature. Linear equations in one, two, and three variables were mastered, in addition to quadratic and biquadratic equations, as well as certain cases of third- and higher-order equations. A great number of geometric problems could be solved, including the area of triangles, regular polygons, and trapezoids; approximative formulas for the area and the perimeter of a circular disk; and the volume of cones and pyramids. The so-called theorems of Pythagoras and Thales were also known and applied in practice, but they were never formulated in a general way; Pythagorean number triples were instead collected in lists. After 1600 B.C. the development of Babylonian mathematics stagnated. Nearly no cuneiform texts on mathematics that originated during the next thousand years have been passed on to us. Only in the second half of the first millennium B.C. do numerous mathematical texts appear once again. This reappearance is probably related to the needs of practical astronomy that developed in this period.

THE GREEKS AND THE CONCEPT OF A MATHEMATICAL PROOF

The Greeks, who settled throughout the Mediterranean region, must have played an important role in preserving and spreading the mathematical knowledge of the Egyptians and Babylonians. It is likely that the first major advance the Greeks made was to consider mathematical concepts (such as numbers and geometric figures) to be abstractions, creations of the human mind, and not part of the real world. A line was no longer considered to be a taut rope or the edge of a pyramid, and a rectangle was not the boundary of a plot of land. Instead, they were defined as ideas, abstractions that physical objects could only approximate. It seems quite certain that the Greek philosophers were the first to realize that a mathematical statement must be proved by logical deduction from certain basic

facts or axioms. Before, mathematical results had been verified by induction—that is, by sufficient experience. It was a most important advance to realize that a mathematical proposition cannot be proved by a thousand or even a million cases for which it is true if, in principle, infinitely many cases may occur.

Thales is given credit for having invented the rigorous mathematical proof. Whether he did or not, it was surely clear to the Greeks that the validity of a given mathematical statement had to be demonstrated. Thales was reportedly a merchant and may have encountered in his travels various formulas for the areas of planar figures and for the volumes of solid objects. Sometimes there would be different formulas for the same area or volume—say, for the volume of the frustum of a pyramid with a rectangular base; the frustum is the truncated pyramid that is cut off at some height h by a plane parallel to the base, as shown in the margin. The Babylonians had one formula for this object, and the Egyptians had another. (The Egyptians certainly had a practical reason for developing accurate geometric formulas: they regularly had to measure their land, because the annual flooding of the Nile River obliterated almost all boundary markers in the flood plain.)

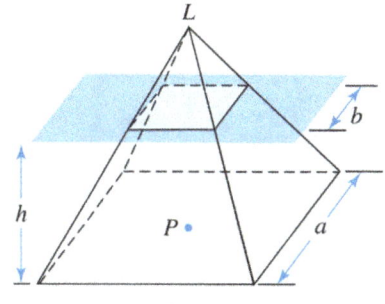

Volume of a frustum of a pyramid with a square base:
$V = \frac{1}{3} h (a^2 + ab + b^2)$.

Now, given that there certainly could not be two correct formulas, Thales faced the problem of deciding first how to find the correct formula, and, second, how to convince others that this formula was valid. Certainly it was not a question for debate in the way that people could debate more subjective questions, such as which was the better of two fine works of art. The method that he evolved for discovering true statements about geometric figures constituted the beginnings of geometry, and the method that he used to convince his associates was probably the origin of the method of proof by deductive reasoning.[1] (Incidentally, the formula the Egyptians developed was the correct one.)

[1] In Chapters 3 and 5 we shall discuss other serious reasons why mathematical statements need a precise proof by reasoning and cannot be deduced by means of an "experimental proof."

Thus, it is not surprising that Thales is said to be the creator of the first Greek geometry. The invention of geometry as an abstract mathematical theory supported by rigorous deductive proofs was certainly one of the turning points of scientific thinking. In particular, it led, as we shall see, to the creation of mathematical models for physical phenomena. For instance, the Greeks had the concept of a curved line, which they defined as the trace of a point moving through space. Obviously, the motion of a planet, if considered a point in the sky, would produce an imaginary curve in the firmament.

CIRCLES, PYTHAGORAS, AND THE MUSIC OF THE SPHERES

The simplest curves are *straight lines* and *circles*. The straight line is the shortest path joining two points in space. It is the trace of a flying insect (thought to be very, very small) that moves in the shortest way from one point to another. A circle is the set of all points in a plane that are at some fixed, equal distance from a center point. If one fixes any point on a sphere and rotates the sphere about a fixed axis, the point will describe a circle.

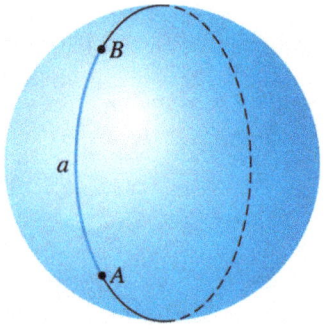

A *sphere* consists of all those points in space that are at some fixed, equal distance from a center point. The shortest connection between two points *A* and *B* on a sphere is an arc *a* of a great circle.

The beautifully symmetric circle does not, at first sight, seem to have a minimum property like the straight line. However, if we rotate a circle about one of its diameters, we generate a sphere. For any two points on the generating circle, the shortest path connecting these points on the sphere must be a subarc of this circle. (Another, very surprising optimum property of the circle also known to the Greeks is connected with "Queen Dido's problem," which we will consider later in this chapter.)

The scientists of antiquity used ideas about circles and spheres to create a mathematical model for the motion of the planets and the stars in the heavens. Pythagoras assumed the stars to be attached to a crystal sphere that revolved daily about an axis through the Earth. Similarly, the seven other ancient planets—the Sun, the Moon, Mercury, Mars, Jupiter, Venus, and Saturn—were each supposedly attached to a moving sphere of their own.

Motion of the celestial sphere.

Somewhat later Plato, in his *Timaios*, linked the creation of time with that of the heavens and the planets. The Sun in its course defines day and night as well as the year, while the Moon describes the month "when on its circular orbit it once again overtakes the Sun," and "The cycles of the other planets men have not understood, except for a few among many, and they neither give names to these cycles, nor do they measure their orbits by numbers, taking observations, so that they truly don't know that the intricate diversity and the marvelous change of the wanderings of those planets also is time." The Pythagorean concept of the planets and their heavens, later developed into a theory for the motion of the heavenly bodies, formed the foundation of astronomy until the sixteenth century. This model was rather complex, and we will mention only one other feature here.

The Pythagoreans believed that numbers were the key to understanding the order of the universe, and to them numbers meant whole numbers (or integers, 1, 2, 3, 4, ...). Aristotle described their beliefs in his *Metaphysics*, saying the following:

> The so-called Pythagoreans applied themselves to mathematics, and . . . came to believe that its principles are the principles of everything. And since *numbers* are by nature first among these principles, they fancied that they could detect in numbers, more than in fire and earth and water, many analogues of what is and what comes to be. . . . And since they saw further that the properties and ratios of the musical scales are based on numbers, that all other things are entirely modeled upon numbers, and that numbers are the ultimate things in the whole physical universe, they assumed the elements of numbers to be the elements of everything, and the whole universe to be harmony and number.

In fact, Pythagoras had discovered a remarkable relation between numbers and musical sounds. For instance, if you pluck a taut string on a guitar, you produce a musical note. The pitch of the note produced depends on the length of the plucked string. Thus a player of a string instrument generates sounds of different pitches by changing the length of the vibrating string. The surprising observation of Pythagoras was that sounds produced by strings are harmonious if their lengths are in the ratio of whole numbers, say 1:2, 2:3, 3:4, or 5:8. In this way, harmony could be explained in terms of integers. To the Pythagoreans this discovery had a mystical significance. They concluded that all relations in nature could be expressed by integers and, accordingly, believed that the ratios of distances between celestial bodies would correspond to the ratios of lengths of harmonious chords. Then the heavenly spheres, as they revolved, produced harmonious sounds that only initiates could hear. This was the *music of the spheres* so often mentioned in literature.

The Pythagorean concept of heavenly harmony was perhaps the first abstract model that attempted to explain complex phenomena in nature by means of a simple and coherent mathematical theory. In fact, Pythagoras has turned out to be so important for the history of mathematics and physics that it is well worth our while to survey his life, legendary though its details may be.

It is reported that Pythagoras was born on the island of Samos, near Asia Minor. He visited Thales, who encouraged him to study in Egypt, the source of much of Thales' original knowledge. Pythagoras

The Pythagorean theorem: in any right-angled triangle, $a^2 + b^2 = c^2$.

traveled in Egypt for many years, acquiring both mathematical and mystical knowledge. When he returned to Samos he founded a religious and philosophical society. For political reasons, Pythagoras eventually had to leave his homeland. He moved to Croton in southern Italy, where he had many supporters and followers.

The society that he founded had a strict set of beliefs and a fixed code of conduct. After a period of testing, the new initiates in the society were permitted to hear the Master's voice, but only from behind a curtain; several years later, when their souls were further purified by the Pythagorean lifestyle, they were permitted to actually see Pythagoras. The Pythagoreans believed that the soul could ascend through the spheres, to eventual union with God, by means of mathematics.

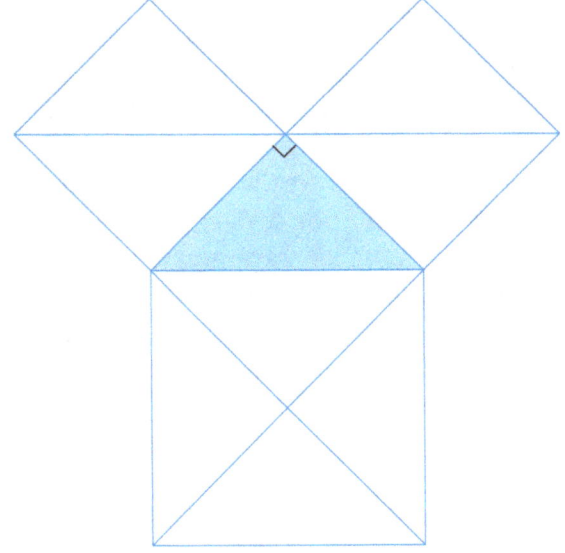

For an isosceles right triangle, it is easy to see by inspection that the sum of the areas of the squares on the shorter sides equals the area of the square on the longer side.

Many mathematical results are attributed to this society—for example, the *Pythagorean theorem*. It states that the sides of a right triangle are related by the formula $a^2 + b^2 = c^2$, where c is the length of the longest side, and a and b are the lengths of the sides that meet perpendicularly. This theorem led the Pythagoreans to discover irrational numbers such as $\sqrt{2}$, the square root of 2, when they tried to deal with a right triangle whose short sides are each equal to 1. The length of its longest side must satisfy the equation $c^2 = 1 + 1 = 2$, whence $c = \sqrt{2}$. From this we conclude that the number $\sqrt{2}$ must exist, because such a triangle exists. On the other hand, it is not difficult to prove that $\sqrt{2}$ cannot be written as a ratio of two integers, that $\sqrt{2}$ is not a rational number. (A number is said to be *rational* if it can be expressed as a *ratio* of two integers, that is, if it is a fraction like 2/3, 17/39, 14/63, ... ; otherwise it is called *irrational*. One can prove that if a number x is rational then its infinite decimal fraction is eventually periodic, such as $2/3 = 0.6666...$, $13/11 = 1.18888...$, or $5/7 = 0.714285714285...$. Moreover, x turns out to be irrational if its decimal representation does not eventually repeat itself periodically. A celebrated irrational number is π, the value of the area of a circular disk of radius 1, as was proved by Lambert in 1761. In fact, Lindemann demonstrated in 1882 that π is a *transcendental* number, that it is not a root of an algebraic equation $a + bx + cx^2 + ... \ dx^n = 0$ with integers $a, b, c, ... , d$ as coefficients.)

The Pythagoreans were shocked to discover the existence of irrational numbers, because these numbers contradicted their belief that everything in the universe could be explained in terms of whole numbers and their ratios. At first they apparently tried to keep the unpleasant truth a secret. The discoverer, Hippasus of Metapontum (about 450 B.C.), as one legend says, was even thrown overboard into the sea to drown after he had made the discovery during a voyage. Perhaps for the first time in scientific history, abstract thinking had led inexorably to a conclusion that totally shattered people's preconceptions. Hippasus possibly discovered that the edge and the diagonal of the regular pentagon are incommensurable—that their quotient is an irrational number. The *pentacle*, the star pentagram in the pentagon (see figure) was the seal of the Pythagoreans. It later

Pentagon

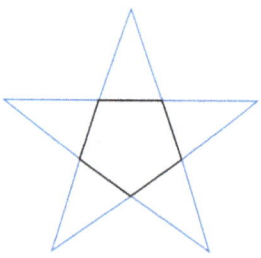

Pentacle

> ### A PROOF THAT $\sqrt{2}$ IS NOT A RATIO OF TWO INTEGERS
>
> To prove this, we will assume that the square root of 2 is a ratio of integers and show that this assumption leads to a contradiction. If $\sqrt{2} = P/Q$, where P and Q are integers, then by dividing out all common factors, we could express $\sqrt{2}$ as p/q, where p and q are positive integers that now have no common integer factors. Suppose that we have $\sqrt{2} = p/q$ with such a pair p and q. If we square both sides of the equation to get $2 = p^2/q^2$ and then multiply both sides by q^2, we have $2q^2 = p^2$. Now let us make a table showing the possibilities for the final digits of $2q^2$ and p^2.
>
If a number ends in	0 1 2 3 4 5 6 7 8 9
> | then its square ends in | 0 1 4 9 6 5 6 9 4 1 |
> | and twice its square ends in | 0 2 8 8 2 0 2 8 8 2 |
>
> Evidently, $2q^2$ must end in one of the digits of the third row, and p^2 must end in a digit of the second row. But the only digit that appears in both rows is 0; so the only way that $2q^2$ can be equal to p^2 is if both of them end in 0. This is possible only if p ends in 0 and q ends in either 0 or 5. In either case, both p and q must be divisible by 5. But this contradicts our assumption that we had divided out all the common factors in P and Q to obtain p and q. This contradiction shows that the original assumption, that $\sqrt{2} = P/Q$, cannot be correct. Therefore, the square root of 2 is not a ratio of two integers.

became a magical symbol capable of averting mischief and harm, an *apotropaion*. Doctor Faustus used it to ban the "Unspeakable One."

According to another report, Hippasus was drowned because he published the discovery that a regular dodecahedron can be inscribed in a sphere, supposedly another secret of the Pythagoreans. This dodecahedron is a convex polyhedron whose twelve faces are congruent equilateral pentagons. It is one of the five "Platonic bodies," probably

The five Platonic bodies. Engraving from Kepler's *Mysterium Cosmographicum*, second edition, Frankfurt 1621. (The first edition appeared in Tübingen, 1596.)

all of them known to the Pythagoreans. The other four are the regular tetrahedron, the cube, the octahedron, and the icosahedron, which symbolize fire, earth, air, and water, respectively. According to Plato's *Timaios*, God used the dodecahedron as a contour of the entire universe. These five bodies that symbolize the cosmos are the cosmic bodies; they appear again in Johannes Kepler's design of the world, published in his *Mysterium Cosmographicum* (1596). Kepler thought he had found the plan God used in creating the cosmos. To this end Kepler set the five regular bodies between the planetary spheres in such a way that any of the cosmic spheres became the circumscribed sphere of one of the cosmic bodies and simultaneously the inscribed sphere of the next higher body. Regrettably, this beautiful idea led to nothing, but it was the beginning of Kepler's great astronomical career and illustrated his speculative way of thought. Contrary to the general opinion of our era of rationalism, many of the great scientists did indeed follow such a speculative (or metaphysical) approach in their scientific investigations.

The secrecy of the Pythagorean society and the mystical observances supposedly learned by Pythagoras in Egypt aroused much suspicion. About 500 B.C., Pythagoras was forced to flee to Tarentum and then to Metapontum, where he was murdered. His followers continued to be active elsewhere until at least 400 B.C.

CYCLOIDS AND CONIC SECTIONS. APOLLONIUS, PTOLEMY, AND THE ALMAGEST

After the dissolution of the Pythagorean school, the study of geometry was continued by many other Greek schools. It is a basic principle of mathematics (as, it seems, of the human mind) to build more and more complicated structures out of simple ones. So from the straight line and the circle, which have such pleasing minimum properties, the Greeks were able to construct more complex curves. For example, let a circle roll on a straight line (in the way that the wheels of a train roll on rails) or on another circle.

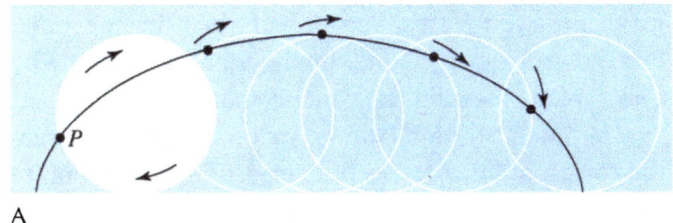
A

Consider the curve described by a point fixed on the periphery of the rolling circle. It is called a *cycloid* if the wheel rolls on a straight rail, and an *epicycloid* if the wheel rolls on the outside of a circular rail, or a *hypocycloid* if it rolls on the inside of a circular rail (see p. 54).

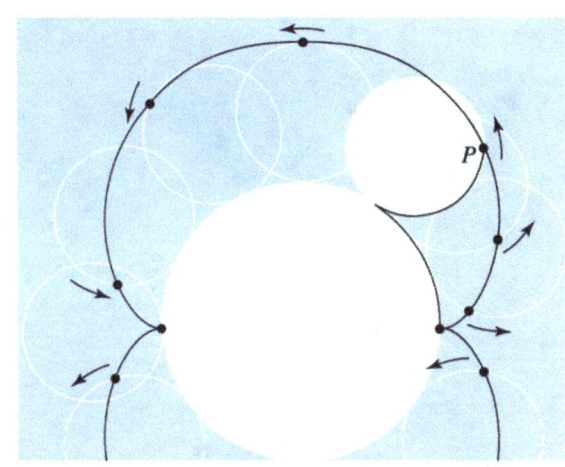
B

Other curves, called *trochoids*, are created by a point attached to a wheel, but not necessarily to its rim, as the wheel rolls on a straight line or on the outside (*epitrochoid*) or inside (*hypotrochoid*) of another circle.

C

D

Roll-curves. (A) Cycloid. (B) Epicycloid. (C) Hypocycloid. (D) Trochoid.

The ellipse as central projection of a circle.

We can produce another kind of curve by projecting a circle along straight lines emanating from a fixed point, the center of projection (the light bulb in the illustration at the right), onto a fixed plane. This curve is an *ellipse*; it is one of the five *conic sections*. As the name implies, a conic section is produced when a plane intersects a cone of revolution. Such a cone can be obtained by rotating a line L through a point P about an axis A, as shown in the illustration on page 54. The

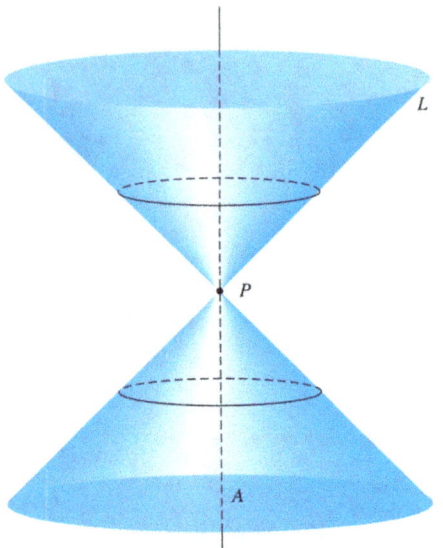

A cone of revolution.

result looks like two ice cream cones with their tips touching. Point *P* is the *vertex* of the cone. Intersection of this cone by planes, each plane meeting the cone in more than one point (as shown in the illustration below), yields the five conic sections: the corresponding curves of intersection, in addition to the ellipse, are the *circle*, the *parabola*, the *hyperbola*, and the *straight line* (not shown in the illustration).

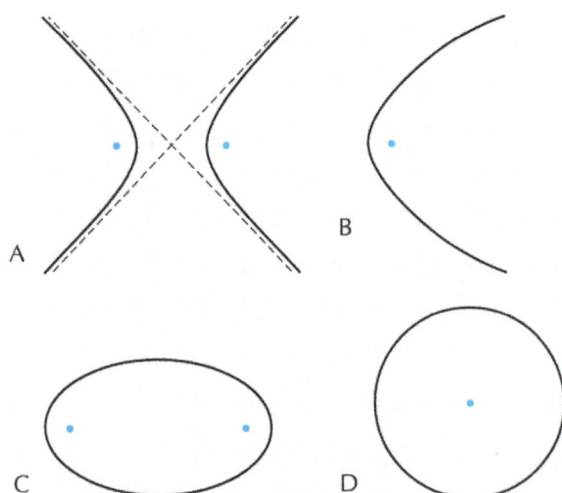

Planar representations of the conic section: (A) hyperbola, with its foci and asymptotes; (B) parabola, with its focus; (C) ellipse, with its foci; (D) circle, with its center.

The conic sections: (A) hyperbola, (B) parabola, (C) ellipse, (D) circle.

Woodcut from Georg von Peurbach's *Theoricae novae planetarum* edited by Oronce Fine as a teaching text for the University of Paris (1515). It was the canonical description of the heavens until the end of the 16th century, and even Copernicus was to a large extent under the influence of this work. Peurbach described the solid sphere representations of Ptolemaic planetary models, which he probably based on Ibn al-Haytham's work "On the configuration of the world" (translated into Latin in the 13th century). The same frontispiece was used for the Sacrobosco edition of the first four books of Euclids's *Elements* (in exerpts) which appeared under the title *Textus de Sphaera* in Paris, 1521.

With the exception of the straight line, each conic section has either one or two focal points: the circle has one focus, its center, as does the parabola; the ellipse and the hyperbola have two foci, located on their major axes. Some of the remarkable properties of these points are described later.

The main results on conic sections were found by Apollonius of Perga (262–190 B.C.). They are described in his eight books, *Conic Sections*, the first four of which were revisions on work done by Euclid, work that was subsequently lost forever. Also, the eighth book of Apollonius's treatise appears to be lost.

The discovery of the conic sections is attributed to Menaechmus, a member of Plato's school of mathematics. This school flourished in the city-state of Athens during the fourth century, just after the passing of the Golden Age of Pericles, the classic period of Greek art, architecture, and philosophy, when the Acropolis was built.

Frontispiece of Vincenzo Viviani's reconstruction of Book V of the *Conic Sections* by Apollonius, with an appendix on "Steiner's problem" (see p. 92).

Frontispiece of Apollonius's *Opera,* translated from Greek into Latin by Joannes Baptista Memus, published by Bernardino Bindoni, Venice, 1537.

Plato, a student of Socrates, founded his school, the Academy, in a sacred area of Athens called Hekademeia (after a hero named Hekademos). All later academies obtained their name from this institution, which existed without interruption for about one thousand years, until it was dissolved by the Emperor Justinian in A.D. 529. Plato's school was something like a small university, where the philosopher and his friends taught courses to students. Two of the greatest mathematicians of antiquity, Eudoxus of Cnidus (408–355 B.C.) and Theatetus (420–367 B.C.), were members of the Academy. Although Plato was not a mathematician, he appreciated this science so much that he required every student to devote ten years to the mathematical sciences, and five more to philosophy. The sign over the entrance to the Academy reputedly read, "*Let no one who is unacquainted with geometry enter here.*"

Ptolemy observing the stars with a quadrant, together with an allegoric Astronomia. (From Gregorius Reisch, *Margarita Philosophica nova*, Strassburg, 1512, an early compendium of philosophy and science.) In those days Ptolemy was often depicted as a king since he was erroneously thought to be descended from the Ptolemaic dynasty, ruling Egypt after Alexander.

It is reported that Plato suggested the following problem to his students: explain the motion of the heavenly bodies by combining various circular and spherical motions. Plato did not particularly appreciate astronomy as a practical science, useful for farmers and seamen alike. In his opinion, it deserved attention only as a playground for the geometer, to whom it was a source of interesting problems.

What problem did the motions of the planets pose? Observed from the Earth, these motions appear to be quite complicated. The motion of the Sun and the Moon can be roughly described as circular with constant speed, but the deviations from the circular orbit were troublesome to the Greeks, and they felt challenged to find some explanations for those irregularities. The observed orbits of the planets are even more complicated, because as the planets go through one revolution, they reverse direction for a time, then reverse again to go forward, all with changing speed. The Greeks sought to understand this seemingly wild motion by a geometric model.

Eudoxus took up this problem and proposed an elegant and purely geometric theory of the celestial motions, which, however, had the serious shortcoming that it could not explain much of the empirical

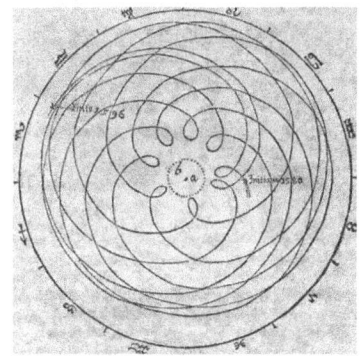

The motion of Mars. From Kepler's *Astronomia Nova* (1609), p. 4.

The apparent motion of the planets. The photograph shows the movements of Mercury, Venus, Mars, Jupiter, and Saturn. The diagram shows the paths traced by the planets as seen from the Earth, which the Ptolemaic theory tried to explain.

data. Thus, in the third century B.C., Apollonius of Perga suggested that the celestial orbits should be explained by means of combinations of circular motions, as the construction of the epicycloid. This idea was to become the most important astronomical theory of the next two thousand years. It was worked out in detail by Hipparchus, the greatest astronomer of antiquity, in the second century B.C. His work is known to us through the celebrated *Mathematical Collection*, written by the Greek astronomer Ptolemy of Alexandria in the second century A.D., and called *al-Magest*, "the greatest," by Arabic-speaking astronomers of the Middle Ages. This book completed the *geocentric system* of astronomy, which later was known as the *Ptolemaic system*. The theory of Ptolemy became so completely accepted that, in the Middle Ages, it was thought to have been handed directly to man by God.

THE HELIOCENTRIC SYSTEM. COPERNICUS, GALILEO, AND KEPLER

It is not surprising that the Greek astronomers placed the Earth and not the Sun at the center of our universe, because we see the Sun rise every morning and sink below the horizon every evening. Still, we may wonder why the Greeks did not at least test the *heliocentric system*, which places the Sun at the center of the universe.[2] In fact, Aristarchus of Samos had conceived such a system in the third century B.C. He taught that the Earth and the other planets move in circular orbits around a fixed Sun. His hypotheses were not accepted for several reasons: the Greeks could not explain why objects would be able to stay on a moving Earth, why people did not sense that the Earth was moving, and why the clouds did not lag behind a moving Earth. These arguments were to be used against the heliocentric theory almost two thousand years later, when it was again proposed by the astronomer Nicolaus Copernicus.

[2]The Pythagorean *Philolaus* (fifth century B.C.) introduced the idea of the motion of the Earth, which together with the Sun, Moon, and the five planets he regarded to be revolving around some fictitious *central fire*. In addition, he introduced a fictitious *Counter-Earth* to bring the number of planetary bodies up to ten, a sacred Pythagorean number.

Copernicus, born in 1473, studied at the University of Cracow but eventually went to Italy to visit Bologna, then a scientific center of Europe. There he learned how to make astronomical observations, a pastime that was to occupy him for the rest of his life. The upshot of his investigations was a truly remarkable book entitled *De Revolutionibus Orbium Coelestium* (On the Revolutions of the Heavenly Orbits). There he introduced his heliocentric theory of the universe that was to bring about the "Copernican revolution" in science. However, he was cautious enough to delay publication of his ideas until he lay near death in 1543. Indeed, *De Revolutionibus* eventually was placed on the papal index of forbidden books (in 1616).

Frontispiece of the *Almagest*. First Latin translation from the Greek by George of Trebizond (Trabzon or Trapezunt), 1528.

Nicolaus Copernicus (1473-1543). Portrait by an unknown painter, about 1600 (Leipzig University Library).

Frontispiece of Nicolaus Copernicus, *De revolutionbus orbium coelestium, libri VI*, Nuremberg, Johannes Petreius, 1543. With a dedication of the printer. This copy was later in the possession of Kepler.

Galileo Galilei (1564-1642)

Galileo Galilei took neither this, nor the burning at the stake of the astronomer Giordano Bruno by the Inquisition in 1600, as seriously as he should have. Ecclesiastical opposition to the ideas of Copernicus became enormous, because these ideas questioned a theory supported by the church authorities. In 1632, Galileo published his *Dialogue Concerning the Two Chief World Systems* debating the pros and cons of the heliocentric theory. This brought him into confrontation with the church and led to his trial, condemnation, recantation, and house arrest at the age of seventy.

As early as 1602, Galileo had discovered that heavy stones fall to the ground as quickly as do light ones. Thus he was led to the mechanical

law that all falling bodies accelerate toward the Earth at the same rate, provided that the resistance of the air could be ignored. His celebrated formula $s = 16t^2$ states that a falling body has traveled the distance s (measured in feet) after a fall of t seconds. He used this discovery to solve an old problem of motion, the problem of finding the trajectory of a ball thrown into the air. Galileo stated and proved that such a ball will travel along a path shaped like a parabola, again disregarding the resistance of the air. This was probably the first time in the history of science that a conic section (other than a circle or straight line) had been used to describe a physical phenomenon.

Yet another scientific revolution was still to come. The German astronomer Johannes Kepler deprived the circle of its right to be the only adequate orbit for a heavenly body! After his studies in Tübingen, Kepler became a professor of mathematics and morals at Graz. In 1600 he came to Prague as an assistant to the famous Danish astronomer Tycho de Brahe, who had collected an enormous amount of new astronomical data that far surpassed that available to the ancient Greeks. After Brahe's death in 1601, Kepler succeeded him as "Imperial Mathematician" to Emperor Rudolf II, who had his residence in Prague. By working with Brahe's observations, Kepler was forced to the conclusion that the planets do not move in circular orbits with the Sun at the center, but in elliptical ones with the Sun at one of the foci. For the first time the ellipse entered astronomy, the very ellipse that had been studied as a purely mathematical object by Apollonius two thousand years earlier.

Galileo Galilei, *Dialogo... sopra i due massimi sistemi del mondo Tolemaico, e Copernicano,* published by G. B. Landini, Florence, 1632. This celebrated defense of the Copernican theory remained on the *Index Librorum Prohibitorum* until 1823. This etching by Stefano della Bella precedes the frontispiece of Galileo's book and depicts Ptolemy (middle) and Copernicus (right).

The law of the elliptic orbits was only the first of three fundamental laws of planetary motion discovered by Kepler. This great astronomer had been convinced of the mathematical design of the universe, but he always checked his theories against observed facts. Although his search for mathematical harmony led him to amazing flights of scientific fantasy, he eventually arrived at his celebrated third law. It states that the square of the time T that a planetary body requires to complete one orbit is proportional to the

Johannes Kepler (1571–1630)

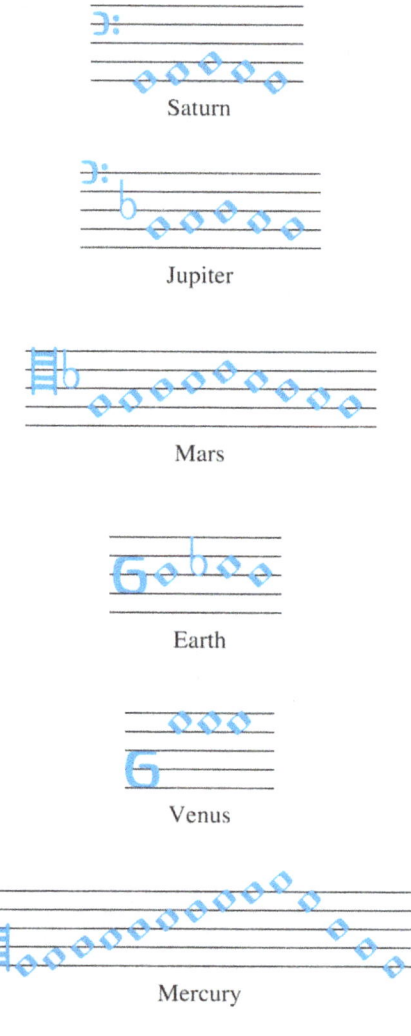

Tycho de Brahe's observatory at Uraniborg, Island of Hven, Denmark. Etching by Ph. Kilian from *Historia Coelestis*, printed by Simon Utzschneider, Augsburg, 1666. This collection of Tycho's vast astronomical observations, much used by Kepler for the Rudolphine tables, remained unpublished for more than 60 years after Tycho's death.

The Pythagorans believed that the heavenly spheres, as they revolved, produced harmonious sounds. Reproduced here is the music of the planets as imagined by Kepler.

cube of the great axis a of its elliptical orbit. In mathematical terms, there is some number K, independent of the planet, such that $T^2 = K \cdot a^3$.

Kepler's second law states that if a planet moves from a point A to some point B in a certain amount of time, and moves from A' to B' in the same amount of time, the area of the sectors SAB and $SA'B'$ must be equal. In other words, "equal areas are swept out in equal times." This explains why planets or comets speed up as they near the Sun. For example, Halley's comet needs 76 years to complete one full orbit; yet it whips around the Sun in only a few days.

The discovery of these beautiful laws, which may be counted among the greatest achievements of the human mind, led Kepler to write in his *Harmony of the World* (1619):

> The wisdom of the Lord is infinite; so also are His glory and His power. Ye heavens, sing His praises! Sun, Moon, and planets, glorify Him in your ineffable language! Celestial harmonies, all ye who comprehend His marvelous works, praise Him. And thou, my soul, praise thy Creator! It is by Him and in Him that all exists. That which we know best is comprised in Him as well as in our vain science. To Him be praise, honor, and glory throughout eternity.

Thus what began with the spheres of Pythagoras and of Eudoxus, and with the circles of Apollonius, Ptolemy, and Copernicus, proceeded to the parabolas of Galileo and ended with the ellipses of Kepler. Isaac Newton based his enormously successful theory of gravitation on Kepler's and Galileo's results. From that theory, Euler derived an *optimum principle* describing the motions of the planets about the Sun, and his work was further developed by Lagrange, Hamilton, and Jacobi. Their work led to *variational principles*, on which all of classical mechanics could be built. The story leads on to atomic physics, quantum mechanics, and general relativity. However, before we see more details, let us return to antiquity.

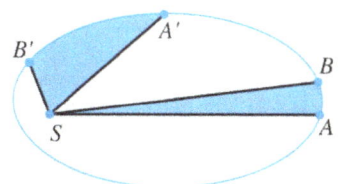

Kepler's second law.

Poem by Johannes Kepler on Copernicus, written in his own hand. Kepler inserted this poem in his copy of *De revolutionibus*, kept at the library of Leipzig University.

THE PROBLEM OF QUEEN DIDO

A celebrated maximum property of the circle already known in antiquity is connected with a story about Queen Dido of Carthage, which was told by the Roman poet Vergil (a contemporary of the Emperor Augustus) in his famous epic *Aeneid*.

Dido was a Phoenician princess from the city of Tyre (now part of Lebanon). She fled by ship from Tyre when King Pygmalion, her ruthless brother, murdered her husband to usurp her possessions. When Dido arrived in Africa about 900 B.C. at the place that later became Carthage, she tried to buy land from the local ruler, King Jarbas of Numidia, so that she and her people could settle and establish a new homeland for themselves. It could be that Dido was thrifty, or that Jarbas did not want any new settlements in his country: they concluded the bargain under the condition that the queen would obtain only as much land as she could enclose by the skin of an ox.

Dido made the most of the situation. First, she interpreted the word "enclose" as broadly as possible. Supposedly she had her people cut the hide in thin strips and tie them together to form a closed cord of great length. This length might have been between 1,000 and 2,000 yards if we assume that the width of the strips was as small as 1/10 inch.

Second, she spread the string on the ground in such a way that it enclosed the largest possible area. By assuming the ground to have been completely flat, we can imagine that Dido had to solve the following mathematical problem: "among all possible closed curves of a given length, find those for which the area of the enclosed inner region is maximal." We may very well suppose that Dido found the correct solution, which is a circle whose circumference is of the prescribed length, and thus acquired an area between 25 and 60 acres. If the two ends of her string had been fixed at two points of a supposedly straight part of the beach of the Mediterranean sea, she would have obtained even more land, by spreading the cord in the form of a semicircle; this is, in fact, what the story says she did. If you look at the maps of medieval European walled cities, you will find that their inhabitants usually came to the same conclusions as Dido.

Dido Purchases Land for the Foundation of Carthage. Engraving by Matthäus Merian the Elder, in *Historische Chronica*, Frankfurt a.M., 1630. Dido's people cut the hide of an ox into thin strips and try to enclose a maximal domain.

Vergil described the foundation of Carthage as follows:

> When they had come to the place where you now will see the mighty walls and the growing castle of the new Carthage, they bargained for the land now called Byrsa, "The Hyde" [named after Dido's transaction], and they got as much as they might enclose by an ox skin.

More details were described by Justinus, a Roman writer from the second or third century A.D., in his *Historiae Philippicae*:

> Elissa [Dido] orders to cut the hide into tiniest strips, and so she acquires a larger piece of land than she had coveted.

Medieval map of Paris.

Medieval map of Cologne.

View of Carthage with Dido and Aeneas, by Claude Lorrain (1600-1682), Hamburger Kunsthalle.

The Greeks were well aware of this mathematical task,[3] called the "isoperimetric problem" (*isos* = equal, *perimetron* = circumference), and they had obvious practical reasons for treating this problem. For example, can the size of an island be guessed from the time that it takes to sail around it? As the illustrations to the right show, clearly one could make terrible mistakes.

Clever people knew that a piece of land having a small periphery might be larger in area than some other piece that has a larger circumference. Thus they could cheat less intelligent owners when exchanging their land, provided that the size of a field was measured by the time it took to walk around it. Such situations were described by Proclus about A.D. 450 in his commentary to Euclid's first book. We will return to the isoperimetric problem later to see that it has been one of the most stimulating and influential problems in the history of mathematics.

Sailing around an island.

Incidentally, Vergil's *Aeneid* is the story of Aeneas of Troy, who, after the fall of his city to King Agamemnon, escaped by ship with some of his followers. He then traveled from Asia Minor through the Mediterranean sea, finally landing in Italy, where he founded the city of Rome. In his travels, he stopped at Carthage, where he met Dido. She fell in love with Aeneas and wanted to marry him. But Jupiter intervened and ordered Aeneas to leave Dido, who then, in desperation, killed herself. For this reason, Dante, in his *Divine Comedy*, condemned her to the second circle of Hell. Henry Purcell, in the days of England's rise to empire, retold this story in his opera *Dido and Aeneas*. Let us listen to the *canto quinto* of Dante's *Inferno*:

[3] The first proof of the isoperimetric property of the circle appears in the commentary of Theon to Ptolemy's *Almagest* and in the collected works of Pappus. The author of the proof is Zenodoros, who must have lived sometime between Archimedes (died 212 B.C.) and Pappus (about A.D. 340), since he quotes Archimedes and is cited by Pappus. However, Zenodoros's proof still contained a gap that was not closed until the second half of the nineteenth century, when the German mathematician Weierstrass provided a complete proof in his lectures at the University of Berlin.

70 THE PARSIMONIOUS UNIVERSE

Aeneas and Dido: from *Codex Vergilius Romanus* 3867 (probably from the sixth century A.D.). Biblioteca Apostolica Romana. This codex is one of very few examples of book paintings from late Roman times.

*Now I begin to hear the sad notes of pain,
now I have come to where
loud cries beat upon my ears.
I have reached a place mute of all light
which roars like the sea in a tempest
when beaten by conflicting winds.
The infernal storm which never stops
drives the spirits in its blast;
whirling and beating, it torments them.
When they come in front of the land slide,
they utter laments, moans, and shrieks;
there they curse the Divine power.
I learned that to such a torment
carnal sinners are condemned
who subject their reason to desire.
And, as starlings are borne by their wings
in the cold season, in a broad and dense flock,
so that blast carries the evil spirits.
Here, there, up, and down, it blows them;
no hope ever comforts them
of rest or even of less pain.
And as cranes go chanting their lays,
making a long line of themselves in the air,
so I saw coming, uttering laments,
shades borne by that strife of winds.
I asked, 'Master, who are these people
whom the black air so punishes?'
'The first of those about whom
you want to know', he said to me,
'was an empress over many peoples,
by the vice of luxury so subdued
that she made lust lawful in her decrees
to take away the blame she had incurred.
She is Semiramis who, we read,
succeeded Ninus, her spouse;
she held the land that now the Sultan rules.
The other is she [Dido] who killed herself
after breaking faith with the ashes of Sichaeus.*

Aeneas leaving Dido. This woodcut illustrates verses 566-607 in Book 4 of Vergil's *Aeneid*. It is taken from the first illustrated print of Vergil, edited by the humanist and poet Sebastian Brant, and printed by J. Grüninger, Strassburg, 1502. This edition was praised as "one of the most wonderful illustrated books ever produced."

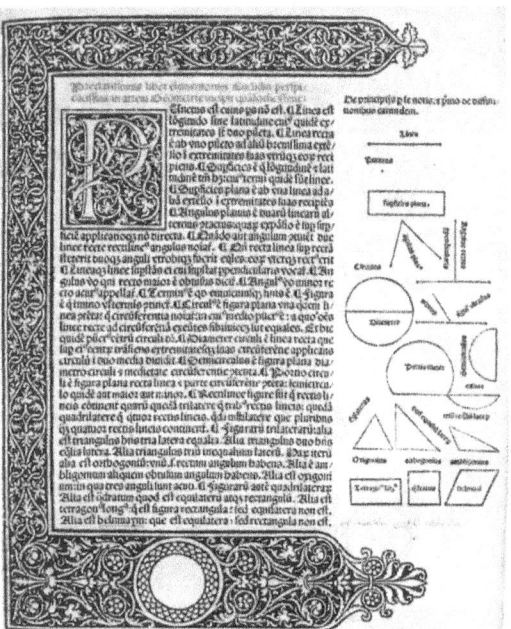

First print of Euclid's *Elements* by Erhard Ratdolt, Venice, 1482.

THE PRINCIPLE OF THE REFLECTION OF LIGHT, AND THE BURNING MIRRORS OF ARCHIMEDES

The first optimum principle in physics explained the reflection of light by a curved mirror.

From the time of Pythagoras, the Greeks had studied light rays. As early as 490 B.C. Empedocles of Agrigentum (in Sicily) stated that light travels through space with finite speed. This truly remarkable insight was not verified until 1676 by the Danish astronomer and mathematician Ole Rømer (1644–1710).

Most of what was known about light in Greek times appears in a work attributed to Euclid (about 330–270 B.C.). He lived in the city of Alexandria, which was founded in 331 B.C. by the emperor Alexander (who was unfortunately called "the Great") at the mouth of the river Nile. Under the dynasty of the Ptolemies it became the scientific center of the world, a position that it maintained for five hundred years. After the fall of the city-state of Athens in 338 B.C., Alexandria became the focal point of Greek culture. In this metropolis, knowledge was collected from and transmitted to the whole ancient world. About 285 B.C., the Museum (or *Museion,* "temple of the Muses") was founded. This was a cultural institution where scholars could work and study. At its peak, the celebrated library of the Museum supposedly contained about 750,000 manuscripts. In Alexandria, Euclid wrote his *Elements* (of mathematics), which became one of the most widespread books in the Western world (about 1,700 editions are known to exist). Actually, the *Elements* consists of thirteen books, in which Euclid collected most of the mathematical knowledge of his age, transformed into a lucid, logically developed masterpiece. The *Elements* is primarily a work on geometry; however, it also contains theorems on number theory and results on areas and volumes. King Ptolemy Soter of Alexandria is said to have asked Euclid if geometry could not be learned by an easier way than by reading the *Elements*. The mathematician's well-known answer was, *"There is no royal road to geometry."*

In addition to the *Elements*, some of Euclid's other writings that have been handed down to us include the *Optics* and the *Catoptrica*[4] ("theory of mirrors"). These books remained until the seventeenth century as basic to geometric optics as the *Elements* was to mathematics.

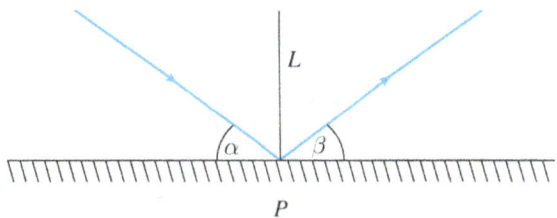

The angle of incidence α equals the angle of reflection β.

Euclid stated in his *Optics* that "light travels through space along straight lines." He then used this and other results to discuss the nature of vision. In the *Catoptrica*, he gave the fundamental *law of reflection*, which states the following two rules for a light ray striking a mirror and being reflected from that mirror:

■ **RULE 1.** The plane of incidence of the ray coincides with its plane of reflection.

■ **RULE 2.** The angle of incidence of the ray equals its angle of reflection.

Assume that the incident ray hits the mirror at point P, and consider the perpendicular L to the mirror at this point. Then the plane that contains the incident ray and the perpendicular is called the *plane of incidence*, whereas the *plane of reflection* is the plane that contains the reflected ray and the perpendicular. Consider the acute angle between the incident ray and the perpendicular. Its complement to a right angle is called the *angle of incidence*. Correspondingly, the complement of the acute angle between the reflected ray and the perpendicular to a 90° angle is the *angle of reflection*.

Then, by rule 1, *the incident and reflected rays lie in the same plane perpendicular to the mirror.* This perpendicular plane intersects the mirror in a planar curve C, which for a flat mirror is a straight line. The construction of the reflected

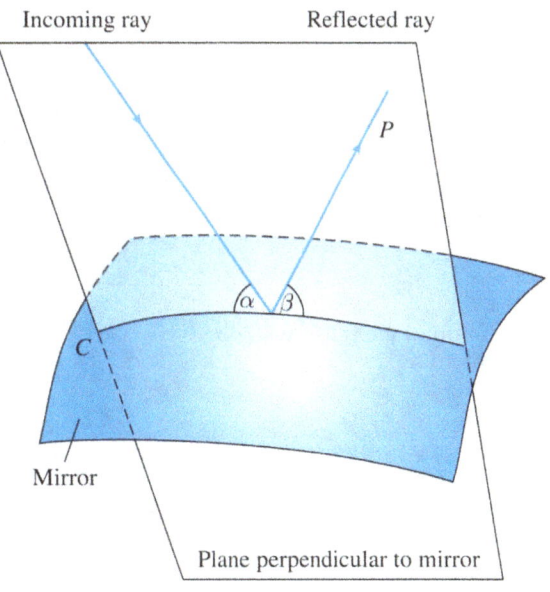

The law of reflection for a curved mirror: $\alpha = \beta$.

[4]The *Catoptrica* is probably not a geniune work of Euclid, but later writers attributed it to him. Hence it is not known who discovered the law of reflection; yet it seems certain that it was known to Archimedes.

What is the shortest path from P to Q via M? That is, what is the shortest path for a cowboy who wants, on his way home, to water his horse?

ray is now a planar problem that can be solved solely in the plane containing the curve C. By rule 2 the angles of incidence and reflection in this plane must be equal. Thus we are led to the *reflection of light rays in planar curves*.

About four hundred years after Euclid, the Alexandrian scientist Heron (about A.D. 100) saw a more fundamental law behind the law of reflection. He stated that this law follows from the principle that *light must always take the shortest path*. To illustrate this, let us consider the reflection from a flat mirror. This problem was seen to be equivalent to the reflection of a ray at a straight line. Hence, let us suppose that on one side of a curve we have two points P and Q, and that we want to travel from P, touch the curve, and then go on to Q. The question then is, what is the shortest path that we can find to accomplish this objective?

Solution of the horse-watering problem.

The same mathematical problem can be posed as a somewhat different "real-world" problem. Suppose that a cowboy out on the prairie wishes at dusk to return to the ranch; yet his horse is very thirsty. The ranch is on the same side of a straight river as the rider. What is the shortest path that the rider can take back home and first water his horse at the river? Heron's answer is that he should travel along a straight line from point P to a point R at the river and then by a straight line from R to Q, where R is chosen so that the angle of incidence equals the angle of reflection. A short proof of Heron's theorem is given in the margin.

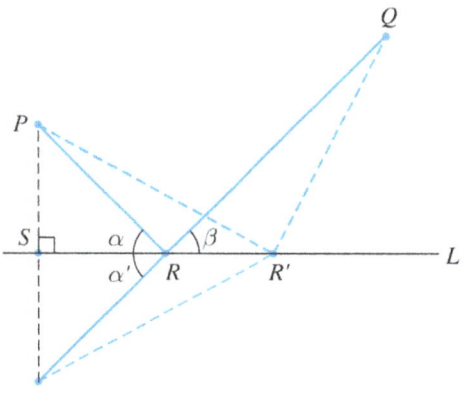

A PROOF OF HERON'S THEOREM
Because a straight line is the shortest connection between two points, we may assume that the shortest path consists of straight lines. Suppose that the shortest path touches line L at some point R. To find the point R, draw the perpendicular from P to L, with the foot on L at S. On this line mark off a distance from S below L equal to the distance \overline{PS}. This determines a point P'. Let R be the intersection point of L with the straight line drawn through P' and Q. Then draw the line from P to R. Note that the three angles α, α', and β at R are equal. Let R' be any other point on L. From the fact that the sum of two sides of a triangle is greater than the third it follows that PRQ is the shortest possible path joining P and Q that touches L.

The question, "What is the shortest path?" is indeed a very natural one. At one time or another, we have all been in a great hurry to reach some place and have wondered how to get there in the shortest possible way. The preceding problem can be generalized in many possible ways; let us look at a few of them. Consider, for example, two points P and Q between two lines L and M. We can ask what the shortest path from P to Q is that first touches line L, then line M, and finally goes to Q. The answer is a broken line that satisfies the law of reflection both at L and at M.

In our first examples we have considered reflections at straight lines, which correspond to reflection by flat mirrors. The Greeks also investigated reflections by curved mirrors or at curved lines, including conic sections.

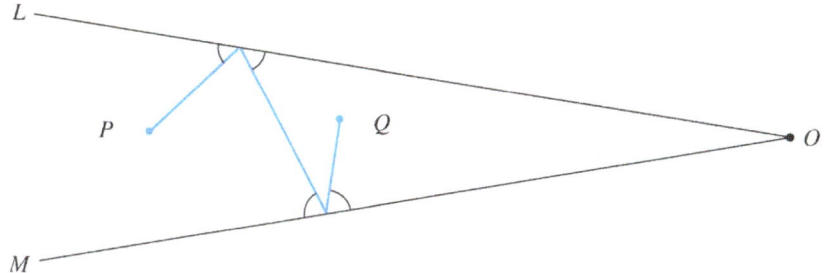

The shortest path between two points P and Q that first touches line L and then line M.

Let us begin with an *ellipse*. As we have already seen, the Greeks knew that every ellipse has two foci, say, P and Q. These foci have a remarkable property:

> For all points R on the ellipse, the sum $\overline{PR} + \overline{QR}$ of the distances of R to P and Q has the same value, say, the value d. Moreover, for any point R in the plane, the value of the sum $\overline{PR} + \overline{QR}$ is smaller than d, larger than d, or equal to d if R lies inside, outside, or on the ellipse, respectively.

Let us now draw the tangent line to the ellipse at some point R as shown in the illustration below. Then the property of the ellipse just stated implies that the broken path from P to R and then on to Q must be the shortest possible path from P to Q by way of the tangent line. Using Heron's result that the minimum-path property implies the law of reflection, we find that the angle of incidence must equal the angle of reflection. This means that the light rays emanating from one focus will be reflected by the ellipse in such a way that all of them pass through the other focus. If we now generate an elliptical mirror by revolving the ellipse about its major axis (which is the straight line through its foci), we obtain a mirror with the following frightening property: if we place an intense light source, such as the Sun, at one of the foci, then its rays will all be collected at the other focus, burning everything found there.

A parabola has similar burning properties. In a way, a parabola can be considered an ellipse in which one of its foci has been moved to infinity. Therefore, all the light rays that emanate from the focus form a bundle of parallel rays after they have been reflected by the

Reflection of light within an ellipse.

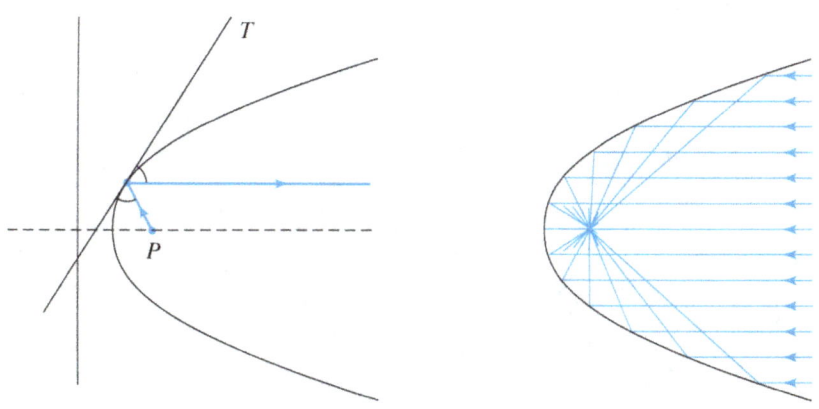

Reflection by a parabola: (A) a parabola with focus P; (B) all parallel rays are reflected through the parabola's focus.

parabola. Imagine the reverse situation—namely, many light rays coming in from a great distance away, as from the Sun. These rays would effectively be parallel to one another; so the parabola would reflect them all, causing them to meet at one point, at the focus of the parabola. Let us rotate the flat parabola about its axis to form a three-dimensional parabolic mirror. Such a mirror would collect all rays parallel to its axis at the focal point. Anything at this focus would be quickly incinerated. For this reason, the Sun could be used as a terrible weapon.[5]

In 216 B.C. the city of Syracuse in Sicily, which was part of Greek civilization, had allied itself with the city of Carthage in its battle against Rome during the second of the three Punic wars. As a result of this coalition, in 214 B.C. the Romans laid siege to Syracuse. Legend tells us that the "burning mirrors" of Archimedes enabled the citizens of this ancient city to use the bright Sicilian summer sun to incinerate the Roman fleet. What were those mirrors?

[5] In general, one cannot focus a mirror on an object to be burned *and*, at the same time, receive the Sun's rays parallel to the axis. Instead the solar rays and the axis of the parabolic mirror will form an angle, and after reflection, the rays will not focus in a single point, but will form a surface called a *caustic* or *burning surface*. This is a very complicated geometric object, but, in a first approximation, it often has a cusp point, and it turns out that most of the rays pass through a minute ball around this cusp. Hence there is a strong concentration of rays near the cusp point, and consequently the burning effect can still be achieved.

Frontispiece of a treatise on optics by the 13th-century Polish scholar Witelo. In 1572, Friedrich Risner added Witelo's work to his edition of the *Opticae Thesaurus*, the famous treatise on optics by the Arabic scientist Al-Haytham (latinized as Alhazen, about 965-1039), who refuted the Greek theories of vision (for example, that visual rays emanate from the observer's eye). Instead, he based his theory of vision upon the action of light, and assumed that light propagates with finite speed in all directions. His theory of the structure and functions of the eye were not surpassed until the 17th century.

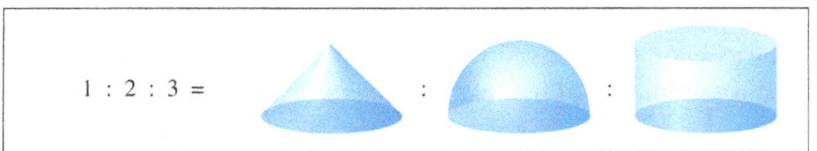

Archimedes's theorem: the ratios of the volumes of a cone, a half-ball, and a cylinder, all of the same height and radius, are 1:2:3.

Archimedes (287–212 B.C.), the greatest mathematician of antiquity, spent most of his life in his native town of Syracuse. It is said that he completed his studies at Alexandria's Museum, but we know for certain only that he visited Egypt on at least one occasion.

The scientific fame of Archimedes is based mainly on his numerous mathematical discoveries. For instance, he found an excellent approximation for the value of π, the area enclosed by a circle of radius 1. He was able to show that π must be a number between 3 1/7 and 3 10/71. Moreover, he calculated many difficult volumes of bodies and areas of surfaces, such as the volume of a ball of radius r ($4\pi r^3/3$) and its surface area ($4\pi r^2$). In fact, he proved that the volumes of a cone, of a half-ball, and of a cylinder, all of the same radius and height, are in the ratio 1:2:3, a theorem of which he was very proud. These mathematical results by Archimedes were truly astounding accomplishments for their time.

Archimedes, in addition to having been an extraordinarily gifted mathematician, was a mechanical genius, as much admired by his contemporaries as by later writers. Among his celebrated inventions were the *Archimedean screw*, a machine for raising water, which was used in Egypt for the irrigation of fields and in Spain for pumping water out of mines, and the *Planetarium*, a set of spheres in which the motion of celestial bodies was imitated by a clever mechanism. Large replicas of such contrivances can be found today in many of the great museums of science. Apparently they were as popular in antiquity as they are now. The writer Claudianus (about A.D. 400) described Jupiter's amazement on seeing his work copied in a glass sphere by a mathematician from Syracuse, and Cicero reported that Marcellus, the conqueror of Syracuse in 212 B.C., took two planetaria as his only booty from the captured city.

Yet the greatest fame of Archimedes in antiquity resulted not from his peaceful inventions but from his contributions to the defense of Syracuse against the Romans. When the Romans besieged the city in 214 B.C., they met an enemy who the 73-year-old mathematician had supplied with unexpected and powerful weapons. The approaching Roman legions were knocked down by a very effective artillery, which discharged long-range stone missiles, and by an enormous number of short-range projectiles. The ships of the Roman fleet were sunk by huge cranes, which either dropped stones on them or lifted their bows with an iron hand and then suddenly smashed them onto the water. Plutarch (in his *Life of Marcellus*) reported the commander Marcellus as saying to his army:

> Shall we not make an end of fighting against this geometric Briareus [a hundred-armed giant of Greek legend], who uses our ships to ladle water from the sea, who has ignominiously threshed and driven off the sambuca [the Roman assaulting ladder, used for the scaling of the walls of a besieged city], and who by the multitude of missiles that he hurls at us all at once outdoes the hundred-armed giants of mythology?

The Roman soldiers were terrified. Plutarch reported,

> If they only saw a rope or a piece of wood extending beyond the walls, they took flight, exclaiming that Archimedes had once again invented a new machine for their destruction.

But for the Roman soldiers, the worst was still to come: the incineration of their ships by the "burning mirrors" of Archimedes. Were such mirrors actually used? They were not described by historians of the day, but appear only in stories written much later. They are mentioned for the first time in a book by Galen (A.D. 129–199), one of the two great physicians of antiquity besides Hippocrates. In 1200 the Byzantine historian Tsetses, unfortunately not a particularly trustworthy writer, gave the following account of what Archimedes supposedly had accomplished:

> When the Roman vessels were within bowshot range, Archimedes made a kind of hexagonal mirror, and small ones with 24 sides, which he placed at an appropriate distance and which he could move with

Woodcut from a treatise by Guiseppe Ceredi which appeared in 1567. Ceredi was interested in the construction and use of the Archimedean screw for the irrigation of fields and the drainage of swamps.

A schematic figure depicting the functioning of an Archimedean screw (from Ozanam's *Dictionnaire Mathématique*, Paris, 1691)

the aid of hinges and metal wires. He arranged the hexagonal mirror so that it was bisected by the meridian of winter and summer, and the sun rays received by this mirror, after being reflected, kindled a great fire which reduced the Roman vessels to ashes, even though they moved out of bow's reach.

Such an array of reflecting plates could have created the same effect as a parabolic mirror. In the sixth century, the mathematician and architect Antheminus also suggested that Archimedes had used a hexagonal mirror. Many historians have dismissed this story as pure myth. The claims and doubts about Archimedes' brilliant accomplishment have stirred up a great deal of controversy for centuries. However, the feasibility of Archimedes' contrivance has been demonstrated experimentally, in 1747 by the French naturalist Georges Buffon and in 1973 by the Greek engineer Ioannis Sakkas.

Death of Archimedes. Mosaic, probably from the school of Raphael. (Städtische Galerie; Liebieghaus Frankfurt am Main.)

The siege of Syracuse lasted two years, and the city finally fell because of treason. In the aftermath of the assault, the old scientist was slain by a Roman soldier, even though the commander Marcellus had asked his men to spare Archimedes' life. As legend goes, Archimedes was sitting in front of his house studying some geometric figures drawn in the sand. When a Roman soldier approached, Archimedes shouted out, *"Don't disturb my figures."* The ruffian, feeling insulted, slew the great man.

Marcellus, as some compensation, erected a tomb for Archimedes on which was depicted a sphere circumscribed by a cylinder, symbolizing, according to Archimedes' wishes, his favorite theorem about the volumes of the cone, cylinder, and sphere. When Cicero (106–43 B.C.) visited Sicily, he was able to locate the monument, overgrown with underbrush and thorns.

Archimedes apparently considered his contributions to engineering to be of minor rank, unworthy of later remembrance. Plutarch reported the following:

> He did not deign to leave behind him any written work in such subjects; he regarded as sordid and ignoble the construction of instruments, and in general every art directed to use and profit, and he only strove after those things which, in their beauty and excellence, remain beyond all contact with the common needs of life.

If we think of all the horror caused by the scientific and engineering inventions used in a multitude of wars, Archimedes' approach appears to be rather reasonable. The great English mathematician G. H. Hardy wrote in 1940

> There is a real mathematics of the real mathematicians, and there is what I will call the "trivial" mathematics, for want of a better word. The trivial mathematics may be justified by arguments . . . , but there is no such defense for the real mathematics, which must be justified as art if it can be justified at all. There is nothing in the least paradoxical or unusual in this view, which is held commonly by mathematicians. . . . There is one comforting conclusion which is easy for a real mathematician. Real mathematics has no effects on war. No one has yet discovered any warlike purpose to be served by the theory of numbers or relativity, and it seems very unlikely that anyone will do so for many years. It is true that there are branches of applied mathematics, such as ballistics and aerodynamics, which have been developed deliberately for war and demand a quite elaborate technique: it is perhaps hard to call them "trivial," but none of them has any claim to rank as "real." They are indeed repulsively ugly and intolerably dull. . . . So a real mathematician has his conscience clear; there is nothing to be set against any value his work may have; mathematics is, as I said at Oxford, a harmless and innocent occupation. The trivial mathematics, on the other hand, has many applications in war. . . . And the general effect of these applications is plain: mathematics facilitates (if not so obviously as physics or chemistry) modern, scientific, "total" war.

Alas, there is no pure, innocent mathematics. Even number theory, the favorite of Hardy, is now a tool of cryptography in the hands of the secret services, not to mention the formula $E = mc^2$ from special relativity theory.

Leonhard Euler (1707–1783). Painting by Emanuel Handmann, 1753. Öffentliche Kunstsammlung Basel.

3 Shortest and Quickest Connections

"Undoubtedly," said Wimsey, "but if you think that this identification is going to make life one grand, sweet song for you, you are mistaken. . . . Since we have devoted a great deal of time and thought to the case on the assumption that it was murder, it's a convenience to know that the assumption is correct."
(Lord Peter Wimsey, in Dorothy L. Sayers's *Have His Carcase*)

At one time or another, we have all considered a problem in the calculus of variations. When planning an automobile trip, we wonder which route is the shortest, or quickest, to our destination. The answers to these two questions are not necessarily the same: it might be quicker to choose a detour on a highway than to take a shorter connection by a small and crooked road that leads through a mountainous area.

Gerardus Mercator (1512–1594), the "cosmograph of the Duke of Jülich," began work at the University of Duisburg in 1552. He became famous for his globes and maps, distributed over all of Europe. His celebrated map of the world, using the new Mercator projection, appeared in 1569; it was mainly meant to serve navigators on sea and is still used today.

Problems of this kind were of interest to the Romans, who connected Italy with the other provinces of their empire by an excellent system of roads. Good roads allowed rapid movements of the Roman legions to suppress any rebellion; yet a rebellion could therefore also very quickly reach Rome itself.

The question of shortest or quickest connections became especially important to the European powers during the fifteenth and sixteenth centuries, when they were searching for the best routes to the Far East and to the New World. Faster sailing routes promised greater profits. The well-known expeditions of Vasco da Gama and of Christopher Columbus must be seen mainly in economic terms.

No wonder that in this period the spherical shape of the Earth again became common knowledge. The early Greeks had imagined the Earth to be a flat disc, surrounded by the ocean and vaulted by the heavens. However, Pythagoras had taught that the Earth was spherical in shape, and Aristotle as well as Archimedes tried to give proofs of this fact. Eratosthenes (about 275–195 B.C.), principal of the Alexandrian library, even derived a fairly accurate value for the circumference of our planet, about 37,000 kilometers. The spherical shape of the Earth was certainly common knowledge to the educated in ancient Greece; yet, in late Roman times, this geographical knowledge was rarely remembered. Most of the Church Fathers inferred from the Bible that the Earth must be a flat disc. Accordingly it was depicted as a wheel having Jerusalem as its center. However, Greek geography was never quite forgotten and, during the Renaissance, Greek writings were rediscovered, and science revived. In 1492, when Christopher Columbus discovered America, Martin Behaim constructed the first globe since antiquity. With the increase of trade and navigation, the preparation of accurate maps became a major scientific task. The cartographer Gerardus Mercator was one of the first to draw charts based on mathematical principles. (It was Gauss who only later understood the basic mathematical principles of map making. We will discuss some of his achievements later.)

The mathematical theory of shortest lines on surfaces, such as on the surface of the Earth, was founded in 1697 by the brothers Jacob and

Johann Bernoulli. Before we look at this theory, we will consider some simpler questions that can be answered by reasoning taken solely from elementary geometry.

Hermann Amandus Schwarz (1843–1921).

THE TRIANGLE OF LEAST PERIMETER INSCRIBED IN AN ACUTE TRIANGLE

The first problem we will examine was posed and solved by Hermann Amandus Schwarz, a professor first at Göttingen and then at Berlin, and one of the most distinguished researchers in the calculus of variations in the nineteenth century. He often enjoyed writing on elementary subjects, and the following problem appears in one of his papers.

> Given an acute triangle, that is, one in which all angles are less than 90°, find an inscribed triangle with the smallest possible perimeter.

By *inscribed triangle* in a given triangle, ABC, we mean a triangle such that each of its vertices P, Q, R touches a different side of ABC.[1]

The answer to the minimum question is not at all obvious. Schwarz discovered that the inscribed triangle of shortest perimeter is given by the *altitude triangle*. The vertices P, Q, R of the altitude triangle are obtained by dropping the perpendicular from each vertex A, B, C to its opposite side. Let us see why this triangle is indeed the solution of Schwarz's problem. From Heron's principle, it follows that any solution must be a *light triangle*. What do we mean by this? If we think of the triangle ABC as a triangular room with mirrors as walls, such an inscribed triangle represents a closed path of travel for a ray of light in the room. Thus, at each vertex of this triangle, the adjacent angles are equal (see the lower illustration at the right). By elementary geometric reasoning, one can demonstrate that the *altitude triangle* is a light triangle and that there is no other light triangle.

Inscribed triangle.

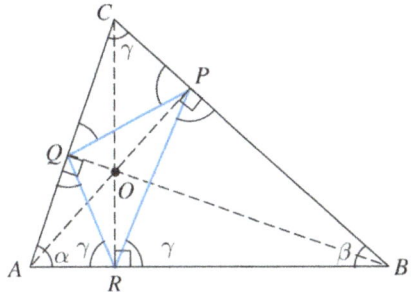

Altitude triangle of the triangle ABP.

[1] We could also ask for the inscribed triangle of largest perimeter. Yet the solution to this problem is completely obvious. It is the original triangle ABC itself.

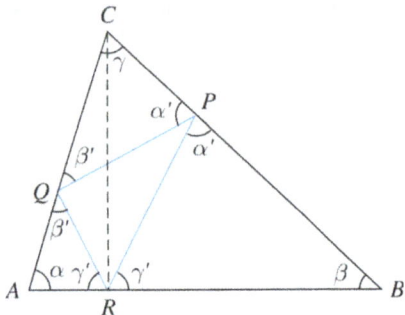

In order to see that the altitude triangle PQR of the triangle ABC is a light triangle, we prove that the angle $\angle ACB = \gamma$ at C is the same as the angles $\angle ARQ$ and $\angle BRP$ at R whence $\angle ARQ = \angle BRP$. From elementary geometry it follows that the altitudes of a triangle meet at a common point O interior to the triangle (see figure at the left). Since $\angle OPB$ and $\angle ORB$ are both right angles, we infer by means of the theorem of Thales that the vertices of the quadrangle $OPBR$ lie on a circle Γ that has OP as a chord passing through its center. We conclude that $\angle PBO = \angle PRO$ since both angles subtend the same arc on Γ. On the other hand, we have $\angle PBO = 90° - \angle ACB$ since the triangle ABQ has a right angle at Q, and $\angle PRO = 90° - \angle PRB$. From these three relations we obtain that $\angle ACB = \angle PRB$. In the same way we see that $\angle ACB = \angle ARQ$, and therefore $\angle ARQ = \angle BRP$. Analogously, we verify that the two angles at P and Q respectively coincide; thus the altitude triangle is indeed a light triangle, as was claimed. Now we want to show that there is no other light triangle than the altitude triangle. So let us consider an arbitrary light triangle $\Delta = PRQ$ in the given triangle ABC. We want to prove that the reflection angle γ' at R coincides with $\gamma = \angle ACB$. Suppose that $\gamma' = \gamma + \delta$. Since $\alpha + \beta + \gamma = 180°$ and $\alpha + \beta' + \gamma' = 180°$ (see figure), we infer that $\beta' = \beta - \delta$, and similarly we reason that $\alpha' = \alpha - \delta$. Inspecting the triangles QPC and ABC, we then conclude $180° = \alpha' + \beta' + \gamma = \alpha + \beta + \gamma - 2\delta$ and $180° = \alpha + \beta + \gamma$, which implies $\delta = 0°$, that is, $\gamma = \gamma'$. Similarly we verify $\alpha = \alpha'$ and $\beta = \beta'$. Now we conclude that the triangle Δ is similar to the altitude triangle since it has the same angles, namely, $180° - 2\alpha$, $180° - 2\beta$, $180° - 2\gamma$, and corresponding edges are parallel. But then it is not difficult to prove that Δ coincides with the altitude triangle.

Summarizing our discussion, we can now state that the altitude triangle is the only light triangle inscribed in ABC.

One might think that, by the preceding arguments, we have outlined a complete proof of Schwarz's theorem; yet we have not. We have proved only that *if there is a solution of Schwarz's problem, it has to be the altitude triangle. But it is by no means clear that there is any solution.* This may look like hair splitting, but, in fact, it is absolutely necessary that

we be meticulous with our proof. In order to convince us of this necessity, let us put ourselves in the position of Lord Peter Wimsey. Suppose that Wimsey has found the dead body of a man and that he has seventeen murder suspects. Moreover, he is completely sure that no one else could be the murderer. By collecting all the evidence and by checking the alibis, he then reduces the number of suspects one by one, and finally only one remains: the butler. Hence the butler is the murderer! But wait. Peter is very careful. By checking the evidence once more, he manages to prove beyond doubt that the dead man died by suicide; so there is no murderer. You see the point: it does not suffice to find a clear and uniquely determined suspect in your criminal case; you also must prove that it is a case of murder.

Karl Weierstrass (1815–1897).

The same principle applies to our mathematical problem. The altitude triangle is our only candidate for a solution of Schwarz's problem, but we also have to prove that a solution exists. You might feel embarrassed if you fell into the trap of thinking that we have a complete proof, but you are in good company, for the greatest mathematicians, including Riemann, made this kind of mistake. Only Karl Weierstrass, professor at the University of Berlin, convinced his contemporaries in 1869 by a striking example that, in order to solve a minimum problem, you must not only track down all the possible candidates for its solution, but also prove that the problem possesses at least one solution.

Here is another example, in case our Lord Peter Wimsey story has not erased all your doubts. We know that 1 is the smallest number among all positive integers 1, 2, 3, 4, 5, … . Assuming that there is a largest integer, we are now going to "prove" that 1 is the largest integer as well. To this end, take any integer n *different from* 1. Then n is smaller than n^2, which is also an integer. Thus n cannot be the largest integer, and therefore 1, being the only remaining possibility, must be the largest of all numbers 1, 2, 3, … . The obvious mistake in this argument is that we assumed the existence of a largest integer.

After the criticism of Weierstrass, mathematicians worked out a kind of reasoning, known as *the direct method in the calculus of variations*,

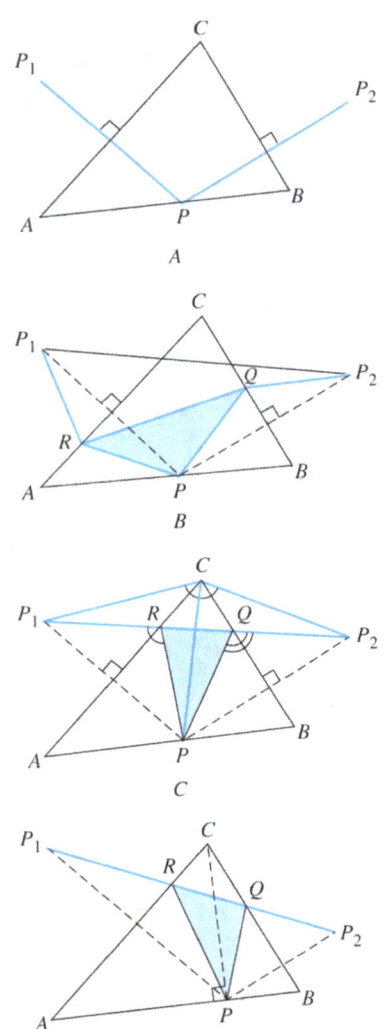

An elementary proof of the existence of a solution to Schwarz's problem.
a. The reflection of an arbitrary point P on the edge AB in the two other sides, AC and BC, of the triangle.
b. The perimeter of an arbitrarily inscribed triangle PQR is, in general, larger than the distance $\overline{P_1P_2}$.
c. For which point P on the edge AB is the distance $\overline{P_1P_2}$ of its two mirror points, P_1 and P_2 smallest?
d. The distance $\overline{P_1P_2}$ is smallest if P is the foot of the altitude from C to side AB.

that often enables them to prove the existence of a solution to a given minimum or maximum problem. For Schwarz's problem, it is, fortunately, fairly easy to prove the existence of a solution by elementary geometric arguments.

To this end, we choose an arbitrary point P on edge AB, as shown at the left, and reflect it in the edges AC and BC. Let P_1 and P_2 be the corresponding mirror images of P. Next, we pick two other points, R and Q, on sides AC and BC, respectively, and consider the inscribed triangle PQR. Its perimeter is equal to the total length L of the line segments P_1RQP_2; that is,

$$L = \overline{P_1R} + \overline{RQ} + \overline{QP_2} = \overline{PR} + \overline{RQ} + \overline{QP}.$$

If we fix P on AB, this length l will be smallest if both Q and R lie on the straight line through P_1 and P_2; so the smallest value of L will be equal to $\overline{P_1P_2}$, the distance between the points P_1 and P_2.

Thus Schwarz's problem is reduced to the question of whether there exists a position of point P on edge AB for which the distance $\overline{P_1P_2}$ is minimal. We claim that this distance is least when P is the foot of the perpendicular from C to side AB. First, note that triangle P_1CP_2 is isosceles because C is at the same distance from the three points P, P_1, P_2; that is, $\overline{P_1C} = \overline{PC} = \overline{P_2C}$. Moreover, the angle of this isosceles triangle P_1CP_2 at C turns out (as you can see from part c of the figure at the left) to be independent of the position of P on the edge AB (it is twice the angle of the triangle ABC at C). An isosceles triangle with a fixed angle subtended by the two equal sides will have the smallest possible base (in this case $\overline{P_1P_2}$) when the length of its two equal sides is minimal, and thus when \overline{PC} is minimal. Because the shortest distance from a point to a line is the perpendicular to that (as we shall see later), it follows that \overline{CP} is minimal if P coincides with the foot of the perpendicular from C to AB.

Hence we have proved that Schwarz's problem possesses a solution and that this solution is furnished by triangle PQR in which P is the foot of the altitude from C to AB (see part d) and Q and R are

intersection points of line P_1P_2 with edges BC and AC, respectively. Because the altitude triangle was the only one that could be a solution, we may infer that *the altitude triangle is the one and only solution of Schwarz's problem*, as we had claimed.

We conclude our remarks on Schwarz's problem with the observation that, besides the light triangle, there are other possible closed paths of light, as the illustration on the right indicates. All follow the law of reflection. They correspond to closed trajectories of a billiard ball on a triangular billiard table. Schwarz's triangle is the only closed path of light that consists of only three straight segments.

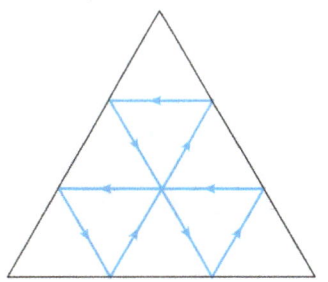

A closed path of light need not be minimal.

As we have seen, it is essential to prove that a solution to a given mathematical problem actually exists. Yet you might think that every "reasonable" mathematical question must have a solution. Here is an example that proves otherwise.

In 1917, the Japanese mathematician Kakeya proposed the following problem:

> Find a planar figure of least area in which a segment of length 1 can be rotated through an angle of 180°, that is, can be completely turned around by continuous movement.

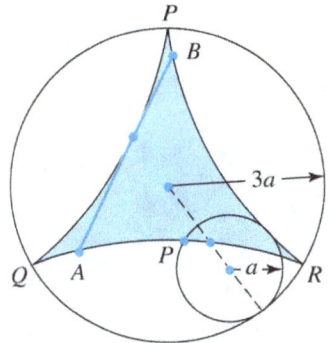

Hypocycloid and needle.

(You might think of a very thin needle that is to be turned around in a flat container, to assume the opposite direction.)

The general "feeling" was that Kakeya's problem should have a solution. For a time, it was believed that a hypocycloid of area $\pi/8$ was the desired figure (see the illustration at the right). However, in 1927, Besicovitch published the following surprising result:

> There are figures of arbitrarily small area in which a segment of length 1 can completely be turned around. Hence one cannot find a figure of minimal area in which the needle can be turned around. Therefore Kakeya's problem has no solution—a complete surprise to anyone who has not seen this question before.

Jacob Steiner (1796–1863).

Evangelista Torricelli (1608–1647). In 1644 appeared his main work, *Opera geometrica,* where he used ideas of Archimedes and Cavalieri preceding the infinitesimal calculus.

STEINER'S PROBLEM

Let us consider another beautiful example of a minimum problem that can be solved by elementary geometry combined with a minimum principle. It was devised by Jakob Steiner, a professor at the University of Berlin, but had already been posed by the French mathematician Pierre de Fermat in 1646. The first to solve this problem was probably Evangelista Torricelli in 1646, Galileo's last student, who is well known for his work on the atmospheric pressure and hydraulics. Similar investigations of the problem were done by Bonaventura Cavalieri (in 1647) and Vincenzo Viviani (in 1659). The latter published his results on what is now called the *Steiner problem* (although the designation *Fermat problem* would be more appropriate) as an appendix to his celebrated reconstruction of the fifth book of Apollonius's *Conic Sections* where the ancient author treated maximum and minimum problems related to conic sections.[2]

In Steiner's problem, three cities A, B, and C are to be connected by a system of roads. We assume that there are no obstructions, so that we are free to build the roads as we please and so that the region around the cities is flat. The problem is to find the system of least possible total length connecting the cities. Then to paraphrase the problem mathematically as follows. We are given three points A, B, and C in a plane and are required to find a point P, and paths joining P with A, B, and C, such that the total length of the paths is a minimum. These paths must be straight since the shortest connection between two points is a straight line. We call any such nodal point P of a minimal connection of A, B, and C a *Fermat point*.

[2]Prior to that, only the first four books of this celebrated treatise were known, and therefore Viviani attempted a reconstruction, based only on some information contained in other ancient writings. Shortly thereafter, an Arabic translation of the first seven books of Apollonius was discovered in the Medicean Library in Florence, a Latin translation of which appeared in 1661. The contemporaries were amazed to see how close Viviani had come to the original. By the way, the eighth book seems to be lost forever.

We would like to find a characterization of Fermat points which shows that there can be at most one such point.[3] For this purpose we consider three different points A, B, and C and assume that P is a Fermat point with respect to A, B, and C. If A, B, and C lie on a straight line and, say, B lies between A and C, then clearly P must coincide with B. Therefore, we can now concentrate on the case where A, B, and C form a true triangle Δ. Let α, β, and γ be the angles at A, B, and C respectively. Then we shall characterize the Fermat point P as follows.

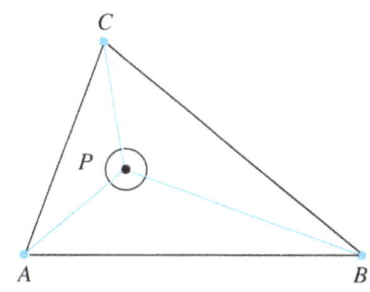

> If all angles in Δ are less than 120°, then P is the uniquely determined point in the interior of Δ from where one sees all edges under the angle of exactly 120°; that is, P is determined by the condition that $\angle APC = \angle CPB = \angle BPA = 120°$. However, if one of the angles in Δ, say, the angle γ at C, is at least 120°, then the Fermat point must coincide with C.

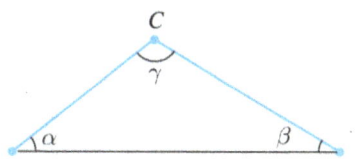

To prove this assertion, we consider an arbitrary Fermat point, for which we distinguish two possibilities:

Case (i): P coincides with one of the corners A, B, and C

Case (ii): P differs from A, B, and C.

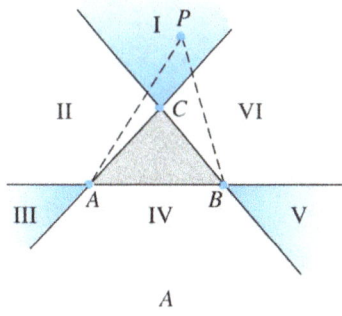

In case (i), the Fermat point P has to be the corner with the largest angle, say, the corner C, because in this case $\overline{AC} + \overline{BC}$ is smaller than $\overline{AB} + \overline{AC}$ and $\overline{AB} + \overline{BC}$.

In case (ii), we conclude that P must lie either within Δ or on one of its edges. If not, we could find a point Q on the boundary of Δ such that the sum $\overline{AQ} + \overline{BQ} + \overline{CQ}$ were less than $\overline{AP} + \overline{BP} + \overline{CP}$ (see figures), and this would contradict the minimum property of the Fermat point P.

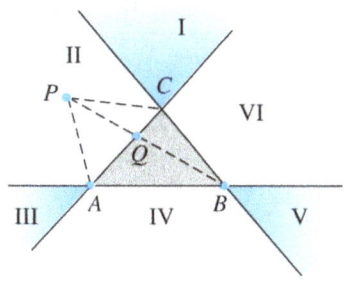

[3] The following mathematical discussion is somewhat technical, but only elementary geometric reasoning is used. It is one of the very few instances where we can show the reader how mathematicians argue. The reader not interested in such details except for looking at the final result described on pp. 97 and 100 should skip this subsection.

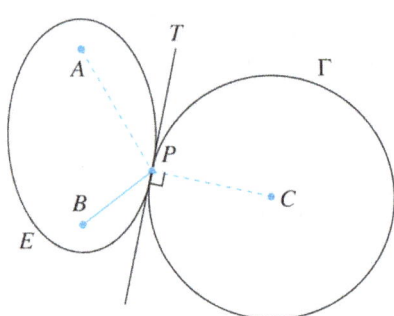

The ellipse E must touch the circle Γ at P tangentially. Therefore PC is perpendicular to E at P, and the reflection principle yields $\sphericalangle BPC = \sphericalangle APC$.

The ellipse E cannot intersect Γ in at least two points since then there are points R on E such that PC is larger than RC while $AP + BP = AR + BR$.

Suppose that P does not lie on the edge AB, and consider the circle Γ centered at P whose radius is \overline{PC}, and the uniquely determined ellipse E with the foci A and B passing through P. If A and B are outside of Γ, E must be tangential to Γ at P, because otherwise we could find a point R on E that lies within Γ, with $\overline{AR} + \overline{BR} + \overline{CR}$ less than $\overline{AP} + \overline{BP} + \overline{CP}$, contradicting the minimum property of P. Let T be the common tangent line of E and Γ at P. Then E lies entirely on the same side of T as A and B. Furthermore, for all points Q on T, $\overline{AQ} + \overline{BQ}$ is larger than $\overline{AP} + \overline{BP}$ as long as Q is different from P. In other words, the broken line APB minimizes the sum of the distances of A and B from T. The reflection law of Chapter 2 then implies that the rays AP and BP meet T at the same angle. Since the ray PC is perpendicular to the line T at P, it follows that $\sphericalangle APC = \sphericalangle BPC$.

We use this result to show first that P cannot lie on one of the edges of the triangle Δ. In fact, if P did lie on the edge AC between A and C, then it cannot lie on the edge AB. Thus we can apply the above result to conclude that $\sphericalangle APC = \sphericalangle BPC$. Since $\sphericalangle APC = 180°$, we obtain that $\sphericalangle BPC = 180°$. This, however, is not true since $\sphericalangle BPC$ is less than $180°$ because B does not lie on the straight line through A and C. Similarly we prove that P does not lie on any of the two other edges of Δ. Thus it follows that the Fermat point P lies in the interior of the triangle Δ.

We have already seen that ∡APC = ∡BPC, and analogously it follows that ∡BPC = ∡APB, that is, the three angles at P are equal, and since they add up to 360°, each angle is 120°. This is *Cavalieri's characterization* of any Fermat point P. Moreover, inspecting once again the figure with the ellipse E, we also realize that the line through PC is the bisector of the angle ∡APB = 120°, that is, it bisects this angle into two equal angles of 60° each.

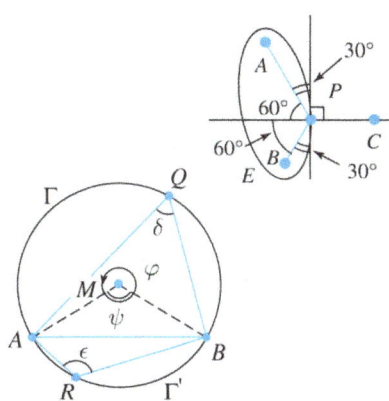

In order to obtain *Torricelli's characterization* of Fermat point P in case (ii), we recall the following generalization of the theorem of Thales: consider two points A and B on a circle K centered at M which divide K into two subarcs Γ and Γ'. Let ϕ and ψ be the two angles under which Γ and Γ' appear from M, that is, $\phi + \psi = 360°$. Finally, let Q and R be points on Γ and Γ', respectively, and set δ = ∡AQB, ϵ = ∡ARB. Then $\psi = 2\delta$, $\phi = 2\epsilon$, and therefore $\delta + \epsilon = 180°$. This means that Γ is a $\psi/2$-arc, that is, *all* triangles ABQ with the edge AB and the opposite corner Q on Γ have the angle δ at Q which is half of the central angle ψ, and similarly Γ' is a $\phi/2$-arc. (The reader can prove this theorem by noting that the triangles AMQ, BMQ, AMR, and BMR are isosceles, and that in any isosceles triangle two of the three angles coincide.)

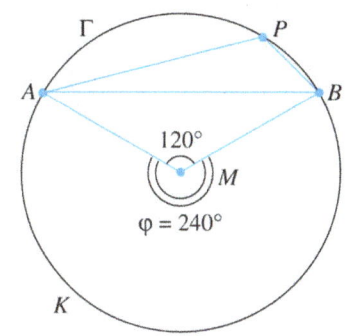

We now apply this result to an edge AB of the triangle Δ = ABC, and to a Fermat point P of Δ. Let K be the circle passing through the points A, B, and P, and centered at some point M. On account of Cavalieri's characterization, we see that P lies on a 120°-arc Γ over the chord AB.

We can construct K from the two points A and B alone, *without using P*, choosing M opposite C so that the triangle ABM over the base AB is isosceles with $\overline{AM} = \overline{BM}$ and ∡AMB = 120°. Then K is the circle centered at M which has the radius $\overline{AM} = \overline{BM}$ (since the complementary angle of 120° is $\phi = 360° - 120° = 240°$, and so the arc Γ on K opposite to M and with end points A and B is, in fact, a 120°-arc, as $\phi/2 = 120°$). We call K constructed in this way the *Torricelli circle* of the edge AB of the triangle Δ. Similarly, we determine the Torricelli circles of the two other edges of Δ, BC and AC, and denote them by K' and K'', respectively. By other elementary geometric reasoning

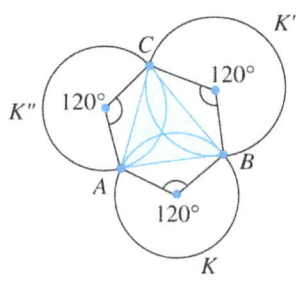

Torricelli circle and the Torricelli point P_0 of ABC.

one can prove that the three Torricelli circles of a triangle always have a common intersection point P_0, the so-called *Torricelli point*. This point P_0 is uniquely determined by Δ, and it is now fairly obvious to realize that P_0 lies inside of Δ if all angles of Δ are less than 120°, and outside of Δ if one of the angles is larger than 120°. If, however, one of the angles (say, γ) is exactly 120°, then P_0 coincides with the corresponding vertex C. Finally, if P_0 lies within Δ, it is, by our previous reasoning, the unique point in Δ such that $\sphericalangle AP_0C = \sphericalangle AP_0B = \sphericalangle BP_0C = 120°$.

Summarizing our results, we see that, for any triangle $\Delta = ABC$ with angles less than 120°, there is at most one Fermat point P within Δ, and if it exists it must coincide with the uniquely determined Torricelli point P_0 lying within Δ. Moreover, if we do not know in advance that all angles of Δ are less than 120° but do know that there is a Fermat point P within Δ, then all angles of Δ are less then 120° and $P = P_0$ is the Torricelli point.

However, it is by no means clear that a Fermat point exists—that there is a solution to the minimum problem—and thus we cannot yet say that the Torricelli point of a triangle (with less than 120° angles) is a Fermat point. All we know at this point is that there cannot be more than one Fermat point P in the interior Δ, since P must coincide with the unique Torricelli point of Δ. We also have seen that no point in the exterior of Δ or on its edges (except for the corner points A, B, C) can be a Fermat point. That still leaves the possibility that there might be a Fermat point P in Δ (which then is the Torricelli point P_0), and at least one further Fermat point on the boundary of Δ, which then is one of the corners, say the vertex A. This, however, is not possible, as we shall presently see. In fact, if there is a Fermat point P within Δ, it has to be the Torricelli point P_0 of Δ, and all angles of Δ have to be less than 120°. Set $a = \overline{AP}$, $b = \overline{BP}$, $c = \overline{CP}$. In order to show that A cannot be a second Fermat point, we prove that $a + b + c$ is less than $\overline{AB} + \overline{AC}$. To this end we first draw straight lines through BP and CP, respectively, and then we draw the perpendiculars from A to these lines, having the feet Q and R. We have $\sphericalangle APQ = \sphericalangle APR = 60°$ since (as we have seen earlier) the lines through BP and CP bisect the 120° angles $\sphericalangle ABP$ and

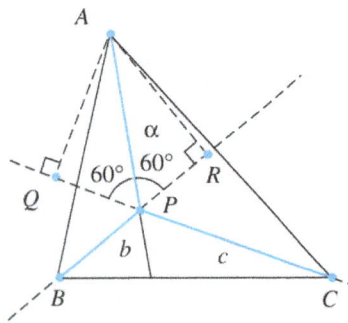

∡APC. Thus, $\overline{PQ} = \overline{PR} = a \cos 60°$ and the cosine of 60° is 1/2. So we have $\overline{PQ} = \overline{PR} = a/2$, and consequently $\overline{QC} = a/2 + c$, $\overline{BR} = b + a/2$. Furthermore, the Pythagorean theorem implies that \overline{AC} is larger than \overline{QC}, and that \overline{AB} is larger than \overline{BR}. Hence it follows that $\overline{AB} + \overline{AC}$ is greater than $(a/2 + c) + (b + a/2) = a + b + c$, as was claimed. Thus, if there is a Fermat point within Δ, the corner A cannot also be Fermat point.

We have arrived at the following result: suppose that there exists a Fermat point P. Then there are only two possibilities:

(1) All Fermat points lie in the interior of Δ. In this case there is exactly one Fermat point, the Torricelli point of Δ; moreover, all angles of Δ are less than 120°.
(2) All Fermat points are corners of Δ.

In case (2) there could, in principle, be one, two, or three Fermat points, but one can prove that there is, in fact, only one such point. This is verified by demonstrating that in case (2) one of the angles of Δ must be at least an angle of 120° whence the corresponding vertex is the unique Fermat point of Δ.

Hence we have the following satisfactory result:

(i) There is at most one Fermat point for each triangle $\Delta = ABC$.
(ii) If there is a Fermat point P, it has to be the Torricelli point P_0 of Δ provided that all angles of Δ are less than 120°; otherwise P has to be the corner with the largest angle.

We still do not know whether or not there is always a Fermat point within a given triangle Δ. With the help of modern mathematics it is in fact fairly easy to prove the existence of such a point, but unfortunately we cannot give an *abstract* existence proof in this book. Later we will offer a reasoning proving existence in a direct way, provided that all angles are less than 120°. If we take the existence of Fermat points for granted, we can finally state a conclusive result for Steiner's problem:

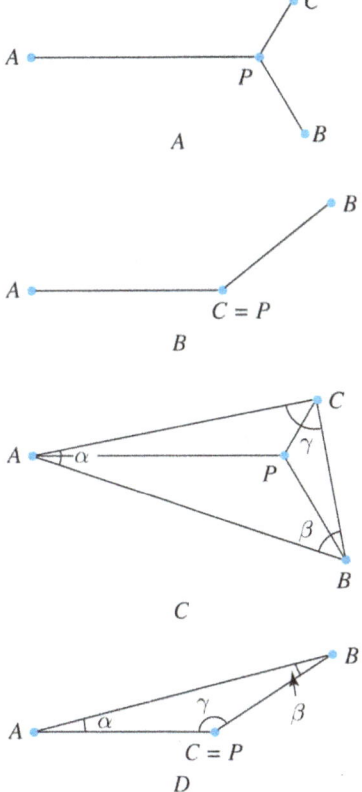

A path of minimal length connecting three points. In case a, all angles in the triangle $\Delta = ABC$ are less than 120°, while in case b the angle at C is at least 120°.

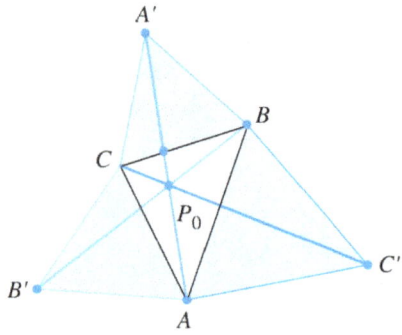

To each triangle $\Delta = ABC$ there is exactly one Fermat point P minimizing the distance sum $\overline{AQ} + \overline{BQ} + \overline{CQ}$ among all points Q of the plane, and P is either the Torricelli point P_0 of Δ or the corner with the largest angle of Δ corresponding to whether or not all angles in Δ are less than 120°.

Now we want to add an elegant construction of the Torricelli point P_0 of a triangle $\Delta = ABC$, found by Viviani (1659) and rediscovered by Thomas Simpson (1750). We consider only the case where all angles of Δ are less than 120°, so that P_0 lies within Δ. First we construct the three equilateral triangles $A'BC$, $C'AB$, and CAB' over the edges BC, AB, and AC which point to the exterior of Δ. Then, according to Viviani, *the Torricelli point P_0 (which in the present case agrees with the Fermat point P) is the common intersection point of the three lines AA', BB', and CC'*. To prove this result, it suffices to show, for instance, that P_0 lies on the line CC', as one can similarly prove that P_0 lies on AA' and also on BB'. We proceed by a surprising "kinematic" reasoning due to Bückner[4]: Consider the shaded triangle ABP_0 in Δ and rotate it by 60° about the hinge B to the exterior. Then P_0 moves to a point P_0^* and A to a point C', and the angle $\sphericalangle BP_0^*C'$ remains 120°. Inspecting the triangle $BP_0P_0^*$, we see that $\sphericalangle P_0BP_0^* = 60°$ (because of the rotation) and is isosceles; hence it has to be equilateral, and therefore all angles in $BP_0P_0^*$ are 60°. Then it follows that the polygon $CP_0P_0^*C'$ is, in fact, a straight line. Finally, the triangle BAC' is also an isosceles triangle since $\overline{AB} = \overline{C'B}$. Since $\sphericalangle ABC' = 60°$, we see that BAC' is in fact equilateral. Therefore, C' is indeed the point described above in Viviani's construction, verifying that it yields the Torricelli point P_0.

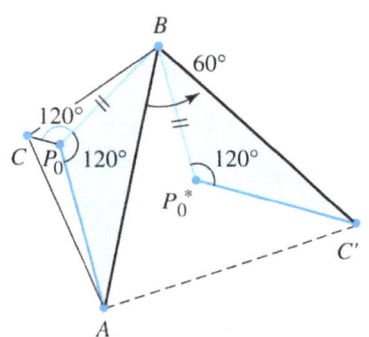

Bückner's proof.

As a bonus to this construction, we note that $\overline{AP_0} + \overline{BP_0} + \overline{CP_0}$ coincides with the distance $\overline{CC'}$, since, as we have seen, $\overline{P_0P_0^*} = \overline{BP_0}$ because $P_0BP_0^*$ is equilateral. The corresponding fact holds true for

[4]This proof, mentioned in the second edition of Rademacher-Toeplitz: *Von Zahlen und Figuren* (1933), was probably found before 1930. A related reasoning was given by T.E. Hofmann in 1929.

$\overline{AA'}$ and $\overline{BB'}$, and so, taking into account that the Fermat point P agrees with the Torricelli point P_0, we obtain the following theorem by F. Heinen (1834):

$$\overline{AP} + \overline{BP} + \overline{CP} = \overline{AA'} = \overline{BB'} = \overline{CC'}.$$

We have therefore found an easy way to determine the minimum value of the distance sum $\overline{AQ} + \overline{BQ} + \overline{CQ}$ among all points Q in the plane, using only a simple geometric construction and no computation.

But we have still not proven that a Fermat point exists. It is now quite remarkable that *Bückner's reasoning can be used to obtain such an existence proof of the Fermat point P for any triangle whose angles are less than 120°*, which simultaneously proves the uniqueness of the Fermat point (within Δ). In view of our previous lengthy discussion, this is even more amazing and shows quite impressively what a striking new idea can achieve, even though it came about three hundred years after the problem was first considered.

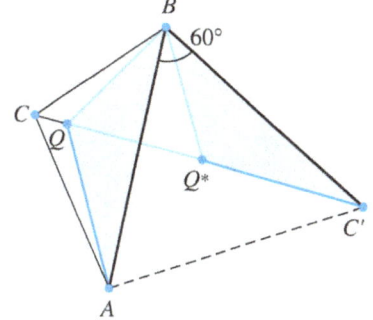

Let us choose an arbitrary point Q in $\Delta = ABC$ and consider the triangle AQB. We rotate AQB by 60° about the hinge B off Δ, whereby Q moves to Q^* and A to C'. We again argue that the triangles QBQ^* and ABC' are equilateral. Then C' is the same point as before and therefore does not depend on the choice of Q. Furthermore, the diagram shows that $\overline{CQ} + \overline{QQ^*} + \overline{Q^*C'} = \overline{AQ} + \overline{BQ} + \overline{CQ}$. Clearly, $\overline{CQ} + \overline{QQ^*} + \overline{Q^*C'}$ is greater than or equal to $\overline{CC'}$ and equals $\overline{CC'}$ only if the polygonal line CQQ^*C' coincides with the straight line CC', which is the case only if $\angle BQ^*C' = 120°$ and $\angle CQB = 120°$, that is, if Q is the Torricelli point P_0 of Δ. Hence, for all Q within Δ, we have $\overline{AQ} + \overline{BQ} + \overline{CQ} \geq \overline{AP_0} + \overline{BP_0} + \overline{CP_0}$, and the equality holds if and only if $Q = P_0$. A close inspection shows that the strict inequality even holds for all points on the boundary of Δ. Furthermore, we have previously shown that, for any point Q in the exterior of Δ, there is a point Q' on the boundary of Δ such that $\overline{AQ'} + \overline{BQ'} + \overline{CQ'} < \overline{AQ} + \overline{BQ} + \overline{CQ}$. Thus by Bückner's reasoning we have proved the following: for any triangle with all angles less than 120° there is exactly one Fermat point P, and this point is the Torricelli point P_0 of Δ.

Fasbender's result.

We leave it to the reader to complete the above reasoning in such a way that we obtain a full proof of: *For any triangle △ there is exactly one Fermat point.*

In 1846 E. Fasbender discovered that a Fermat point P of $△ = ABC$ can also be characterized by a *maximum* principle if P lies in the interior of $△$. Fasbender's result, given without proof, reads as follows:

> The minimum value of the distance sum $\overline{AQ} + \overline{BQ} + \overline{CQ}$ for a fixed triangle $△ = ABC$ is equal to the maximum of the altitudes h of all equilateral triangles circumscribing $△$. If P is the Fermat point, this maximum value is assumed for that equilateral which has the lines AP, BP, and CP, as perpendiculars to its edges passing through A, B, and C, respectively (see figure).

It is of great interest to us that for an interior Fermat point P, equal angles of 120° must be formed by the connecting lines about P. Later, when we treat soap films, soap bubbles, and cracks in certain materials, the 120° angle will occur repeatedly. Moreover, we shall see that soap films can be used as "analogue computers" to obtain a physical solution to Steiner's problem as well as to its generalizations, which we discuss next.

GENERALIZED STEINER PROBLEMS

There is a more general version of Steiner's problem that is of considerable intellectual as well as practical interest. In such a *generalized Steiner problem* we are given a certain number of points in a plane which are to be connected by a system of lines of smallest total length. For example, we can pose the problem of connecting four corners of a square, A, B, C, D, by a *minimal system of paths*. It can be proved that the problem has exactly two solutions (see figure), which can be transformed into each other by a suitable rotation. Consequently, we have to abandon the idea that one can always expect to have exactly one solution, as in the case of Steiner's problem for three points A, B, and C. In the case of many points, the uniqueness of solutions to the problem of finding minimal connecting systems is not the general rule but a fortunate exception. We can speak of *the shortest connection* of two or

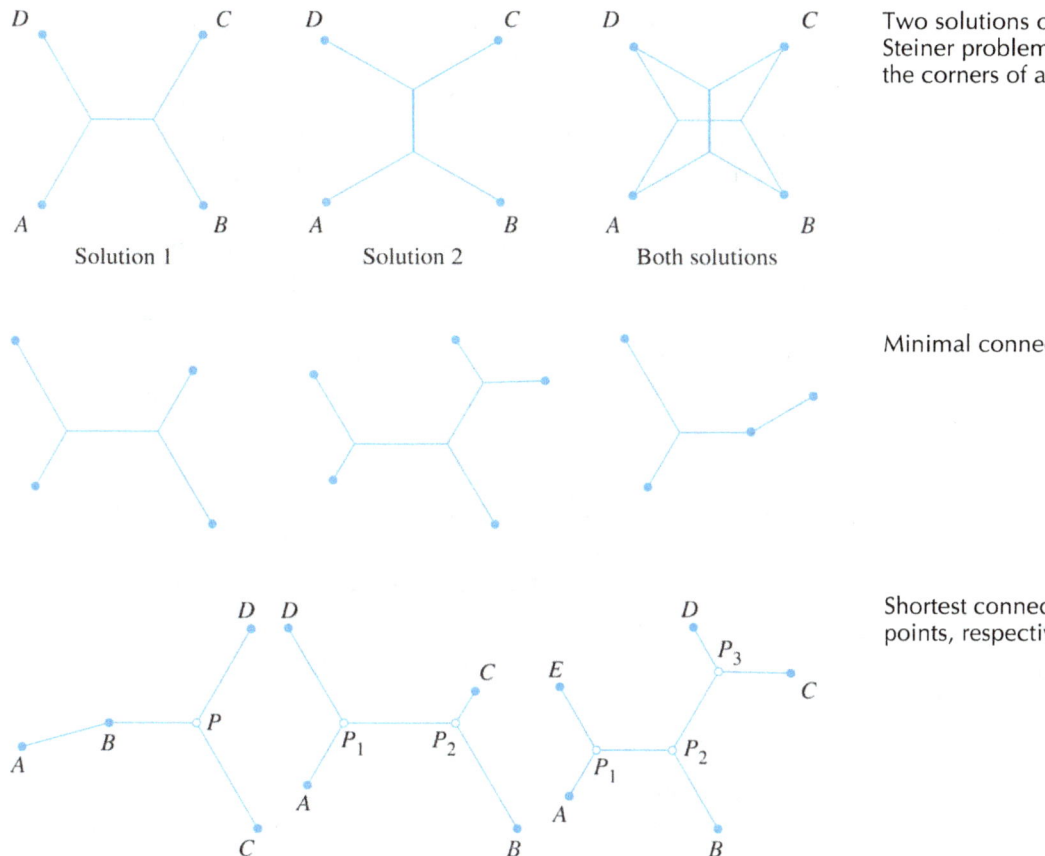

Two solutions of the generalized Steiner problem to four points in the corners of a square.

Minimal connections of four points.

Shortest connections of four and five points, respectively.

three points, whereas 4, 5, 6,... points will in general have several shortest connections; therefore we must be careful to speak of *a shortest connection* between a set of points.

Let us now look at shortest connections of four or five points in a general position (see figures). Such minimal connections look like combinations of the two possible solutions found in the ordinary Steiner problem treated earlier. For example, in the case of four points A, B, C, and D, we have added either one or two extra points P or P_1, P_2, respectively, in order to construct the shortest connection of A, B, C, and D, and for the five points A, B, C, D, and E in our diagram we have added another three points to build the minimal system. Given a set of points, it is *a priori* unclear how many points

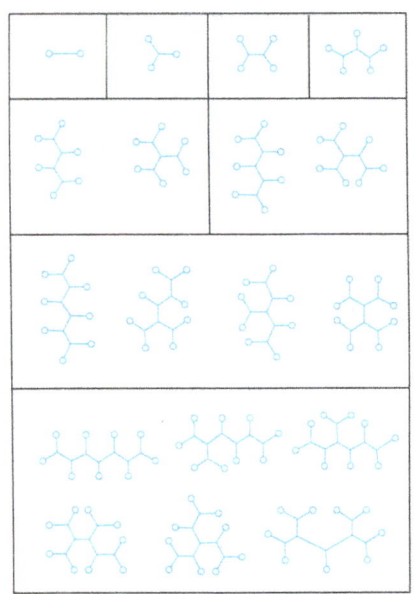

Torricelli trees connecting n points $A_1,..., A_n$, with the maximal number of $k = n - 2$ auxiliary points $P_1,..., P_k$ ($n = 2, 3,..., 9$).

$P_1, P_2,...$ one has to add in order to construct the minimal system of paths connecting the given points, but one can prove that no more than $n - 2$ extra points are needed to build the minimal connection of n given points. Furthermore, as a consequence of the minimum property, at each *extra point* $P_1, P_2,...$ in a minimal connection of given points exactly three lines meet, each pair of which forms an angle of 120°. In other words, extra points in a shortest connection system behave like *Torricelli points* in the ordinary Steiner problem.

Given n points $A_1, A_2,..., A_n$ in the plane, a system S of straight paths connecting the points $A_1,..., A_n$ is called a *Torricelli system spanning these points* if it has the following properties:

(i) There exist k extra points $P_1, P_2,..., P_k$ different from $A_1,..., A_n$, denoted as *nodal points* of S, such that the boundary points of each path either are nodal points or belong to the points $A_1,..., A_n$.

(ii) Each nodal point P_j is a Torricelli point of S, that is, at P_j there meet exactly three paths of S mutually enclosing 120° angles.

(iii) At each of the points $A_1,..., A_n$, one, two, or three paths of S meet. Moreover, if three paths meet at A_j, it must be a Torricelli point of S. If exactly two paths l and l' meet at A_j, then both angles enclosed by l and l' are greater than or equal to 120°.

In the figures on p. 105 we show four different Torricelli systems spanning the same set of 31 points $A_1,..., A_{31}$. (The points $A_1,..., A_{31}$ are drawn completely black, while the associated nodal points $Q_1, Q_2,...$ are indicated by small circles.) There are 12, 13, 14, and 15 nodal points, respectively.

The above discussion (together with a few more or less obvious arguments) shows that each minimal connecting system of points $A_1,..., A_n$ is a Torricelli system, spanning $A_1,..., A_n$. But not each Torricelli system is necessarily a minimal connection, as one can see by inspecting the following six diagrams.

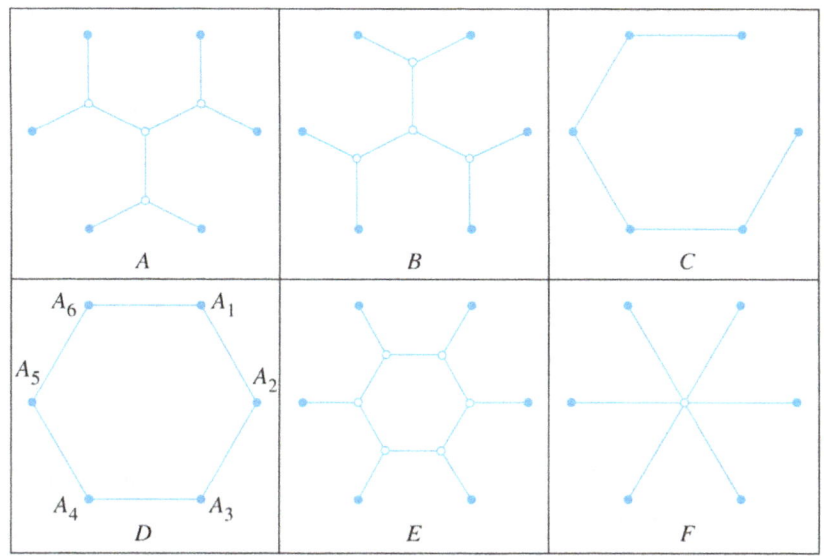

Here A_1,\ldots, A_6 are chosen as corners of a regular hexagon where each side has length 1 (which can be built from six congruent equilateral triangles, a remark useful for explicit computations). All systems in a–e are Torricelli systems, whereas the system in f is not Torricelli, since six lines meet at the central nodal point. The total length of the Torricelli system c is 5, while the total lengths of the two congruent systems a and b is $3\sqrt{3}$, a number greater than 5.15, and so the total length of c is smaller than that of a or b. Moreover, there are five other Torricelli systems congruent to c; each of these six congruent systems is obtained by omitting one edge from the hexagon d. Note that the path systems a, b, and c are even trees, while d and e are not trees. (A path system is, for obvious reasons, called a *tree* if it does not contain a closed loop, that is, if one cannot go in a "circle" within the system.) Note that a Torricelli system can never furnish a shortest connection if it contains a loop, because one could remove one edge of this loop and the resulting system would be shorter. Hence *each shortest connection has to be a Torricelli tree*. The Torricelli systems d and e are obviously not trees, and so they cannot be minimizing connections. We also remark that d and f are limit cases of e, as one obtains f if one lets the loop shrink to the center point, while d is achieved by blowing up the inner loop to the big hexagon. We note

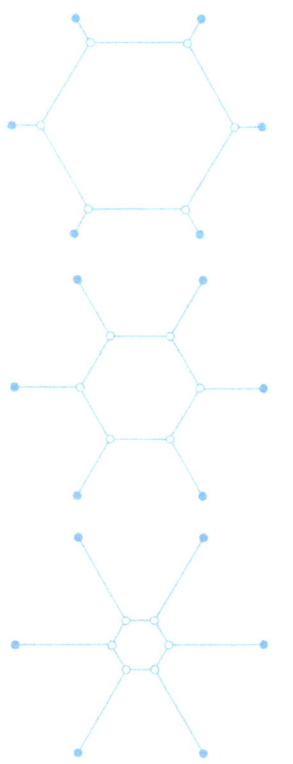

Left: A tree connecting 49 cities in the US. Right: A Torricelli tree connecting the same cities. The Torricelli tree on the right is about 3.1 % shorter than the tree on the left (constructed by Prim in 1957), which is a minimal spanning tree without extra nodal points.

that all members of the family of Torricelli systems e have the same total length, which is the same as the lengths of d and f, even though f is not a Torricelli system but just a limit of such.

For the six corners of the regular hexagon we have found eight different Torricelli trees, namely the six types c and the two types a and b. One can, with some effort, prove that there are no other Torricelli trees, and so the six types c are the six possible shortest connections of $A_1,...,A_6$. The two Torricelli trees a and b are strict local minimizers of the total length since "small changes" of such a tree increase the length; therefore, we call these trees strictly stable.

A shortest connection of given points $A_1,...,A_n$ is said to be a *Steiner tree* spanning these points, because it solves a generalized Steiner problem and is known to be a tree.[5] Our above discussion shows that each Steiner tree is a Torricelli tree while the converse is, in general, not true. Thus, to determine all Steiner trees we have to find all Torricelli trees, and then we have to compare their lengths in order to single out the shortest ones. Unfortunately, this leads to a rather difficult problem of computer science. In fact, Garey and Johnson[6] have

[5]Here we deviate from the notation that is presently used in the literature where our Steiner trees usually as denoted as "minimal Steiner trees." Also, our terminology of Torricelli systems and Torricelli trees is not standard.

[6]M. R. Garey and D. S. Johnson: *Computers and Intractability: A Guide to the Theory of NP-Completeness*. W. H. Freeman, San Francisco, 1979.

Torricelli trees connecting 31 points, generated by soap film experiments.

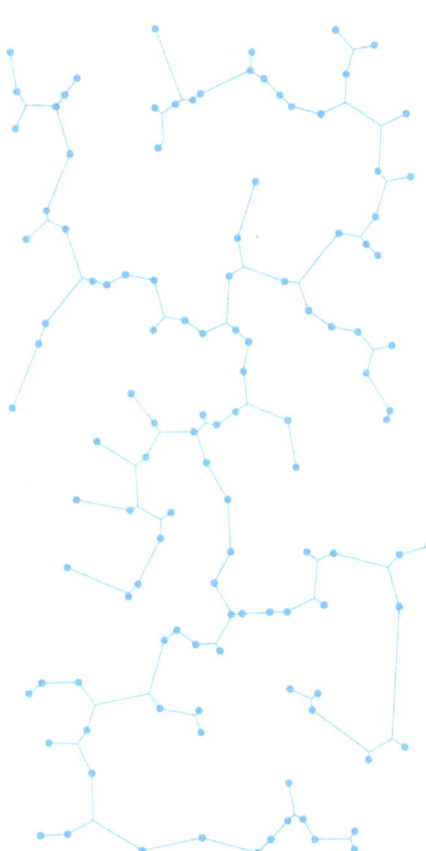

A Torricelli tree connecting 120 points, with 47 extra nodal points, determined by numerical computations.

proved that the solution of the generalized Steiner problem is at least as difficult as any of the so-called NP-complete problems (which are the notoriously unpleasant problems of computer science), as the number of possible Torricelli trees for points in a general position might grow enormously with increasing n. This is rather unfortunate because numerous practical questions can be interpreted as generalized Steiner problems. One might, for instance, think of connecting a number of cities by a road system, a system of telephone lines, gas pipes, or high-power lines, etc. Clearly, one wants to minimize the length of such a system as this very likely will minimize the expenses to build and to maintain it. As another example, a company with co-operating factories distributed over several locations will be interested in minimizing transportation costs; thus it will try to achieve a minimal connection of its factories by a Steiner tree if it can freely build the connecting roads (which should hardly be the case these days).

Another important application of Steiner problems and related mathematical questions is found in the design of computer chips. For example, a logic chip presently carries up to eight million transistors on an area of about one square centimeter to be connected according to their functions by a delicate system of extremely thin wires. Obviously, one wants to keep these wire systems as short as possible in order to pack more and more transistors onto chips, thereby increasing their efficiency. This requires the solution of problems closely related to Steiner problems. For example, the VSLI layout (i.e., the chip design) requires the solution of so-called *Manhattan–Steiner problems* where one tries to find shortest connections of N given points by path systems with only horizontal and vertical branches (see figures).

Biologists use Steiner trees to compute plausible phylogenetic trees of species and then to compare the relationship of species by looking at sequences of amino acids in proteins common to those species being compared. Admittedly, such guesses seem to be rather far-fetched.

The first scientists to pose the generalized Steiner problem and to look at Steiner and Torricelli trees were probably French mathematicians. In the *Annales de Mathematique Gergonne* of 1810–1811 one finds several contributions, some of them anonymous ("par un

Shortest and Quickest Connections

A modern computer chip contains several millions of transistors in an area of just one square centimeter, which are connected in a pattern of horizontal and vertical lines (see p. 108 for two photos showing sections of the chip ZORA). The elementary circuit elements are prestructured as cells. Then the entire circuit plan of the chip is constructed using these cells as components. Our figure depicts the placement plot of the cells of the chip ZORA (40-fold magnification) showing its intricate design.

Two sections of the routing plan for the computer chip ZORA in different magnifications.

abonné"); possibly Gergonne himself was among the authors. Another contribution with practical applications was made in 1827 by Clapeyron and Lamé and can be found in volume 10 of the *Journal of Transportation of St. Petersburg* (in Russian). Also Carl Friedrich Gauss (1777–1855, Professor at Göttingen), thought to be one of the greatest mathematicians of all times, looked at the problem. He contributed to almost all fields in pure and applied mathematics as well as to physics, astronomy, and geodesy. Many of his discoveries are true landmarks in the history of science. In an 1836 letter to his colleague and friend, the astronomer Schumacher, he mentioned a Steiner problem in connection with the task of building a railway system of shortest length between the four cities of Harburg (close to Hamburg), Bremen, Hanover, and Braunschweig; he even thought of posing this problem as a prize question to the students of Göttingen University. But it was only in 1879 that K. Bopp, a student at Göttingen, gave a complete solution of the general Steiner problem for four cities (i.e., $n = 4$). The articles that have been written on the generalized Steiner problem[7] are nearly countless.

MORE ON SHORTEST CONNECTIONS

Another type of minimum problem is to find the shortest path between a point and a curve, or the shortest path between two curves. By the Pythagorean theorem, the shortest connection from a point P to a straight line L can be found by dropping the *perpendicular* onto L. The perpendicular is the straight line through P that intersects L in a right angle at some point P^*.

The shortest path from a point to a line.

From this fact one can deduce that the shortest connection from a point P to a closed convex curve C (where P is supposed to be inside C) will meet this curve at some point P^* at a 90° angle. Just draw the tangent line L to C at P^*; the segment PP^* will then yield the shortest

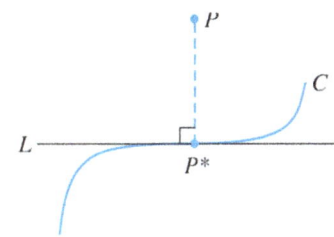

The shortest path from a point to a smooth curve.

[7]The terminology "Steiner problem" was probably coined by Courant and Robbins in their popular book *What Is Mathematics?* Steiner, by the way, only treated the case $n = 3$.

The shortest path from a point to a closed, convex curve.

distance from P to L, and so each of the angles at P^* enclosed by L and PP^* will be a right angle.

Finally, we note that even for an arbitrary smooth curve C, the shortest connection between some point P and the curve will meet C in a right angle; but a correct proof of this fact needs some infinitesimal calculus.

Similarly, the shortest path joining two curves C_1 and C_2 meets both of them in a right angle. Recall what we noticed in the Prologue: soap films, when free to move on other surfaces, always meet them perpendicularly, and cracks in certain materials intersect each other at 90° angles.

SHORTEST LINES ON SURFACES, AND GEODESICS

In the following discussion, we will see perpendicularity play an important role in another, 300-year-old, problem of shortest connections that we stated at the beginning of this chapter, named the problem of finding the shortest path between two points on a curved surface. This has been, and remains, one of the fundamental problems in geometry.

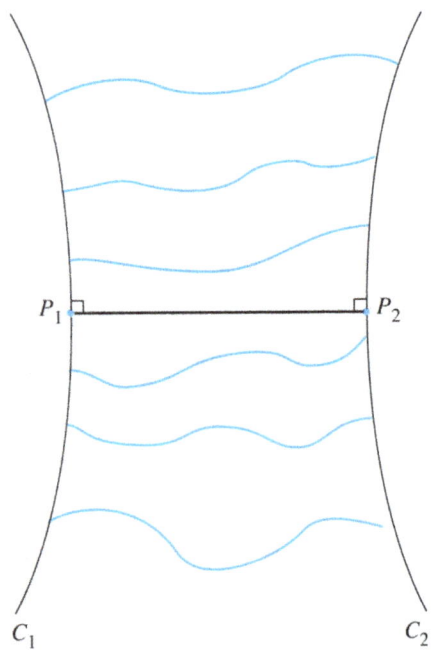

The shortest path joining two smooth curves C_1 and C_2 meets both of them at a right angle.

As we saw, the Greeks knew that the shortest line between two points in a plane is always a straight line and also that light in a homogeneous (isotropic) medium always takes the shortest (or at least locally shortest) path.

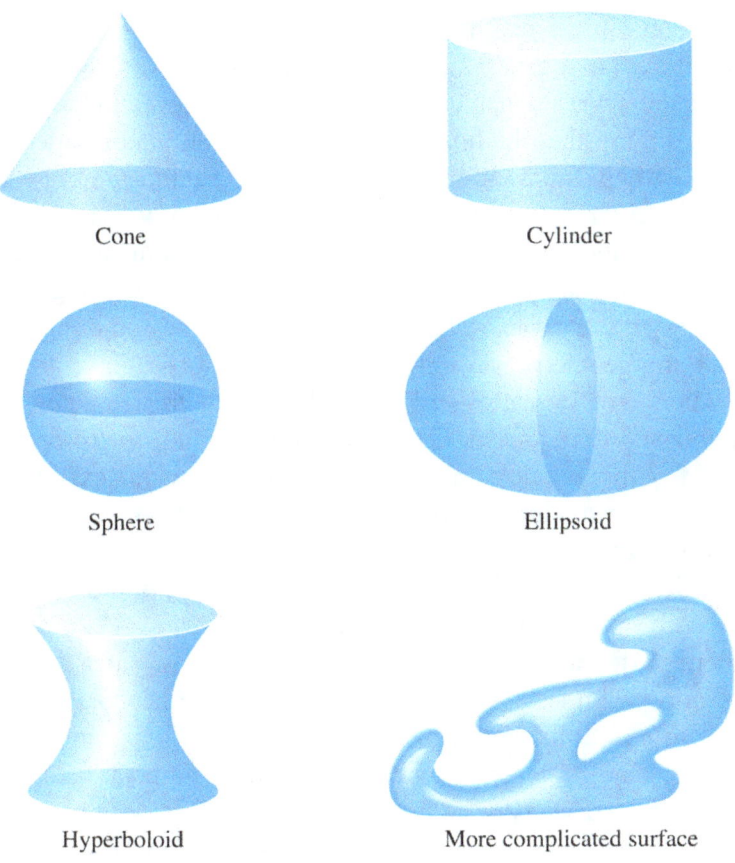

Cone

Cylinder

Sphere

Ellipsoid

Hyperboloid

More complicated surface

What are the shortest connections of any two points on these surfaces? Can one characterize them by geometrical conditions, or, in other words, what properties distinguish the shortest curves from arbitrary ones?

On a curved surface, however, there are generally no straight lines. Even for simple surfaces such as the sphere, cylinder, and cone, it is not immediately obvious what the shortest paths are, and for more complicated surfaces it is much more difficult to obtain an answer.

Before mathematicians could attack this question, they had to have a clear idea of what, in general, a curve and a surface should be. Certainly they are not objects of the physical world, but merely idealizations of such objects, similar to those created by the ancient Greeks. Roughly speaking, we can say that curves and surfaces in mathematics are idealized models of those objects that in daily life are denoted as curves and surfaces, and it is always important to distinguish between the "concrete objects" and their "mathematical models," a distinction that often disappears in a fog of confusion.

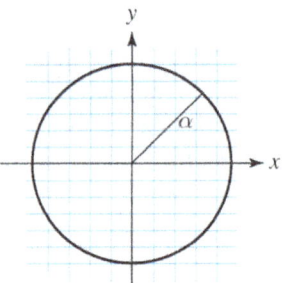

The equation $x^2 + y^2 = a^2$ describes a circle of radius a centered at the point $x = 0$, $y = 0$.

The ancient Greeks had a hard time defining objects like "curves" and "surfaces" in a general way since their algebra was not well developed and always remained on a rather modest level. In fact, some historians think that the final stagnation of Greek mathematics was caused by the Greek's failure to develop algebra and to apply it to geometry. Their approach to geometry was mainly *synthetic* and much less *analytic*; the latter approach is greatly facilitated by algebra, that is, by the technique of a *symbolic computation* using letters or other symbols instead of special numbers. This combination of algebra and geometry was not achieved until the seventeenth century, when René Descartes and Pierre de Fermat created a powerful mathematical tool, *analytic geometry*, which expresses geometry in terms of numbers and equations. A planar curve, for example, is described by a single equation relating the variables x and y, which are plotted against each other on the familiar square grid. For example, the formula $x^2 + y^2 = a^2$ describes a circle of radius a centered at the point $x = 0$, $y = 0$. Subsequently, it was not a great problem to define the concepts "length of a curve," "area of a surface," and "volume of a body" by means of infinitesimal calculus and to compute the length, area, and volume of many interesting geometric objects. More or less the same definitions are still used and are studied every year by multitudes of first-year university students. It is not necessary here to study the technical mathematical details, but one should at least be aware of the fact that, without the development of precise mathematical concepts and flexible and powerful techniques, no progress in mathematics was, nor will ever be, possible. Therefore, the readers of our book as well as of other popular treatises will be able to understand what mathematics is all about but will most likely not be able to do mathematics without studying the necessary details, as "there is no royal road to mathematics."

The mathematical definitions of curves and surfaces, of length, area, tangent lines, tangent planes, curvature, and so forth were developed during the late seventeenth and early eighteenth centuries principally by Cavalieri, Newton, the brothers Bernoulli, Leibniz, and, in particular, Euler, but many other mathematicians participated in these developments.

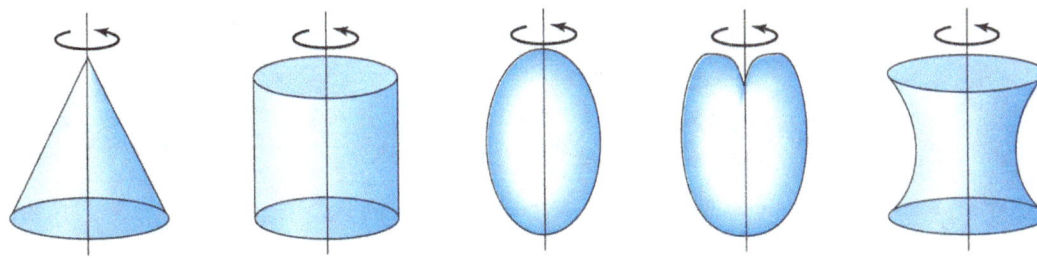

Various surfaces of revolution.

We now begin our discussion of shortest paths on surfaces by considering three surfaces of rotation: the cylinder, the cone, and the sphere.

The cylinder surface contains straight lines—its *meridians* (the lines that generate it as a surface of revolution)—and they certainly are lines of shortest connection. Let us explain the important notion of a meridian. Imagine that a plane curve C, situated on one side of an axis A, is rotated about A, thereby generating a surface of revolution, as shown in the figure. The various positions of C during this process are called the "meridians" of the surface of revolution. They are the generating curves of this surface. For instance, in this way a semicircle generates a sphere, a straight line parallel to the axis of rotation generates a cylinder, and half an ellipse produces an ellipsoid. Other possibilities are depicted above.

For a cylinder, any pair of points P and Q on a meridian has a straight line as the shortest path of connection on the cylinder surface S. What can we say if the points P and Q do *not* lie on the same meridian? Actually, this is not a real problem. We just remember that we may unroll each cylinder into a plane; this is what every house painter does when using a roller. For our purposes, it is somewhat simpler if we imagine S as the skin of a sausage that we cut above P and below Q (see the illustration below) by two planes that meet the cylinder axis perpendicularly. The piece S' that we are left with will be sliced along a suitable meridian M. Now we may spread the sliced skin without any distortion into a plane, where it takes the shape of a rectangle, R. Thus every line on S' is carried over into some line on R, and vice versa.

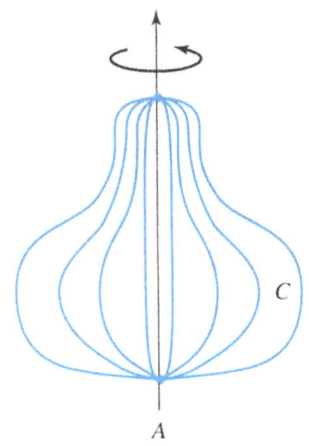

A meridian generating a surface of revolution.

The shortest connection between two points on a cylinder.

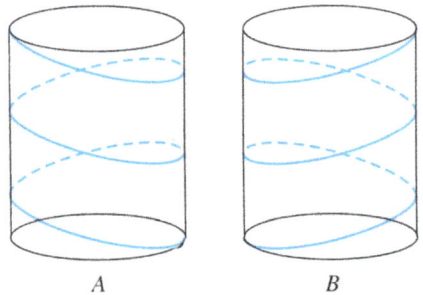

Helices: (A) left-handed; (B) right-handed.

Because we did not distort the skin S' when spreading it into the plane, we have not changed the length of any curve on S'; hence this mapping of S' onto R is called an *isometric mapping*. For this reason, a shortest path on S' corresponds to a shortest path on R. Yet all shortest paths on R are known to us: they are straight lines. If we now reverse our operation and roll R, like a peel, back over the sausage, the straight lines are transformed into helices or circles (which are crushed helices) or straight lines. Thus, the shortest connection of P and Q on S is part of a helix—that is, of the line on a screw, as shown at the left. This is a curve that occurs quite often in nature.

We find a very similar situation if we study the shortest lines on a cone. Notice that each cone can be unrolled onto a planar domain; in fact, by cutting it along a meridian, we can spread it without distortion onto a portion of a plane. Again we employ the fact that the shortest lines in a plane are straight lines. This will immediately give us the shortest lines on S. We need only to roll the planar image of S back onto the cone.

We may proceed similarly if we want to find the shortest lines on a polyhedron. By slicing it at some of its edges, we can again spread it out onto a plane without distortion.

However, the situation is quite different for most surfaces, including the sphere. Just try to spread any part of a sphere—say, the rubber hull of a ball—onto a planar domain. You will realize that one cannot do so without stretching the rubber, causing a distortion.

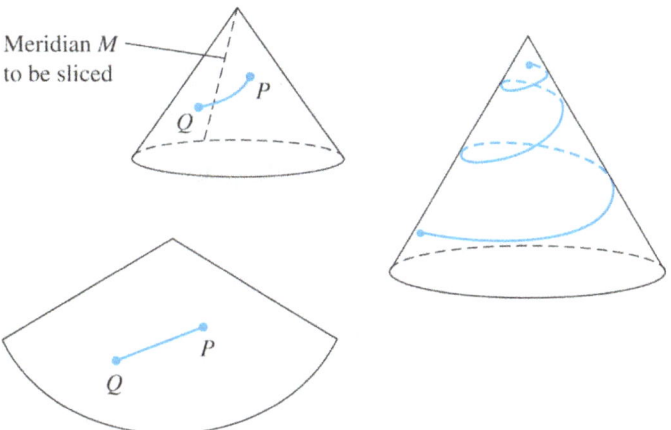

Shortest line on a cone.

There is a mathematical proof that no part of the sphere can be mapped onto a plane without distortion. This proof is based on a concept of the curvature of a surface introduced by Karl Friedrich Gauss, now called the *Gaussian curvature*. This curvature measures precisely how curved a surface is at each of its points. (A definition of this notion will be given in Chapter 5.) Gauss's celebrated theorem then states that we can compute this curvature if we know the length of each curve on the surface. A somewhat more precise statement of Gauss's result is that the metric (that is, the measurement of length) determines the Gaussian curvature of the surface.

To see that a sphere cannot be mapped onto a planar domain without distortion, we compute the Gaussian curvature of a sphere of radius R to be of the same value, namely, $1/R^2$, at each of its points, whereas that of a planar domain is everywhere zero. If there were such a length-preserving map, it would follow that $1/R^2 = 0$, which is, of course, absurd.

Gauss called this important result *theorema egregium* (that is, remarkable theorem). It is the central result of his *Disquisitiones circa Superficies Curvas* (Discourse on curved surfaces), published in 1827, and it is valued as one of the most important discoveries

Carl Friedrich Gauss (1777–1855).

Two pictures explaining Clairaut's theorem.

in mathematics. It explains why we cannot make perfect maps (or charts) of the surface of our planet: the search for maps on which lengths are uniformly proportional to the lengths on the globe is futile. Each type of projection used for chart making will have some disadvantages. That leaves us to choose between the various kinds of errors that the map will have. Yet there is one property that maps can have. Map makers can find representations of the globe in which angles between lines (if not their lengths) are preserved. The Mercator projection provides just such a *conformal map*.

Even though unrolling will not work for the sphere, we can nevertheless find all its shortest lines. They must be portions of great circles on the sphere, because the sphere is symmetric in every direction and each meridian of a surface of revolution furnishes a shortest connection between any two of its points.

For general surfaces of revolution the French mathematician A. C. Clairaut (1713–1765), a member of Maupertuis's Lapland expedition, found the following remarkable result:

> The product of the radius r of the circle of latitude and the sine of the inclination angle α of any shortest, C, with respect to the intersecting meridian has a constant value along C, that is, $r \sin \alpha = $ const for any point.

The first paper on the shortest line between two points *on a general surface* was published by Euler. This paper appeared under the title "De Linea Brevissima in Superficie Quacunque Duo Quaelibet Puncta Jungente" ("On the Shortest Line on an Arbitrary Surface Connecting Any Two Points Whatsoever") in the *Commentarii of the Imperial Academy of St. Petersburg*, in the 1728 issue, which, however, was not publicly distributed until 1732.[8] Euler stated that we can easily solve

 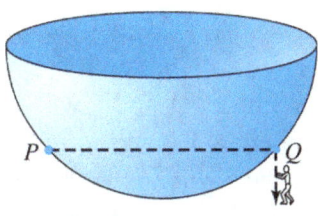

A tight string between two points on a convex and a concave surface.

 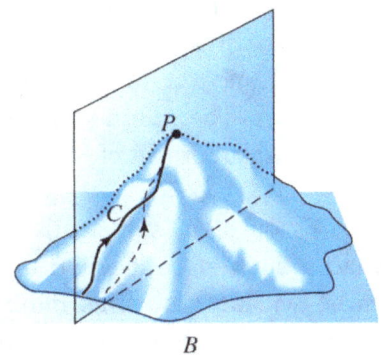

 A *B*

(A) The tangent plane at P and a geodesic C through P.
(B) The osculating plane at P for the geodesic curve C. The dotted curve shows the intersection of this plane with the surface.

the problem of shortest connection between two points on a convex surface by a simple mechanical artifice: we fix a string at one of the points and pull it taut in the direction of the other (see the left-hand drawing in the upper illustration on the next page). The string then yields the shortest connection between the two points.

Obviously this method does not work for a concave surface; so Euler developed a new technique for finding the shortest line. He reduced the problem to the solution of a differential equation. This equation is equivalent to a geometric theorem derived (but not published) by Johann Bernoulli in 1698, in which he stated that *at each point P of a shortest line C, the corresponding osculating plane of C intersects the tangent plane to the surface at P in a right angle* (see the middle illustration).

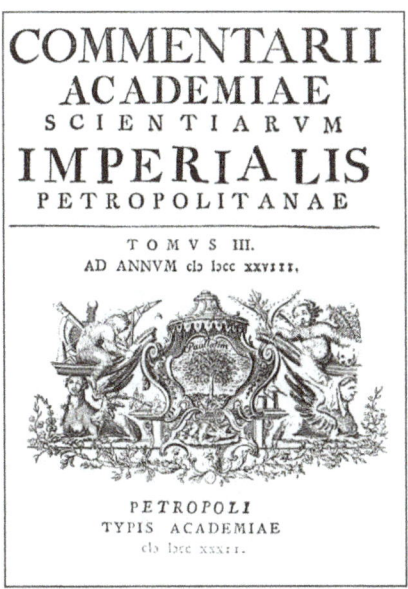

Frontispiece of *Commentarii Academiae Scientiarum Imperialis Petropolitanae*, vol. 3 (1732) and Euler's paper.

[8]For some time, historians believed this paper was incorrectly dated and Euler could not have submitted it for publication before 1729. Recent investigations seem, however, to indicate that the paper may have been written by the end of 1728.

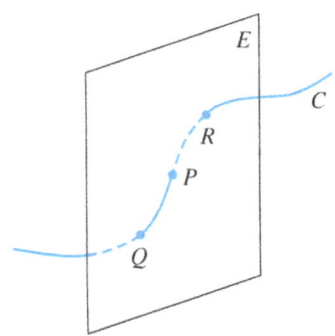

Three points *P, Q,* and *R* on a curve determine a plane *E* that approximates the osculating plane.

What is this *osculating plane* that appears in Bernoulli's theorem? Take any two other points, *Q* and *R*, on *C* close to *P*. In general, the three points *P, Q,* and *R* will determine a plane. Of course, this plane will depend on *Q* and *R*; but if these points both move on *C* toward *P*, the plane will approach a limiting position, and this is called the *osculating plane to C at point P*.

Later on, the name *geodesic line*—or, briefly, *geodesic*—was given to all curves *C* on a surface *S* for which at each of their points *P*, the osculating plane to *C* and the tangent plane to *S* are perpendicular to each other. Then Bernoulli's theorem states that *curves of shortest length must be portions of geodesic lines.*

It turns out that "sufficiently small" portions of a geodesic always yield a shortest connection of their end points; in general, "large parts" will not have this minimizing property. For instance, on a sphere, all the geodesics are parts of great circles. Then, for any pair of points *P* and *Q* that are not at opposite ends of a diameter of the sphere (like the North and South Poles on the Earth), there is exactly one great circle passing through *P* and *Q*. The two points therefore decompose the circle into two arcs, of which the shorter one is the shortest line between *P* and *Q* and the other is geodesic but not a shortest connection (see figure below). The same figure also shows the points *P* and *Q* on a cylinder and on a cone; both pairs are connected by a shortest line, as well as by a geodesic that is not minimal in length.

Complete surfaces, such as the sphere, the ellipsoid, and the cylinder, are defined as surfaces on which you can move with constant speed forever on every geodesic without ever leaving the surface. You

Two points on a surface can sometimes be connected by more than one geodesic.

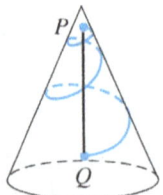

might return to your point of departure, as on a sphere, or you might never return, as when you move along a helix on a cylinder. We have to emphasize the fact that Euler did not prove the existence of a shortest connection between two points on a complete surface. He gave conditions that must be satisfied by such a line. The existence of a shortest connection was not proved until 1900, by David Hilbert.

The study of geodesics had a turbulent beginning. In August 1697, Johann Bernoulli, publicly posed the problem of finding the shortest line between two given points on a convex surface. This was meant as a challenge to his brother Jacob, with whom he was openly feuding. The unfortunate rivalry between the two brothers eventually became so intense, and their polemics so ugly, that the scientific journals of the time declined to publish their results. Let us look at the background of this feud, one of many well-known scientific rivalries.

Between 1666 and 1680, Newton and Leibniz had discovered infinitesimal calculus. It has apparently been proved that both found this powerful mathematical technique independently. However, Newton's method was not published until 1711, whereas Leibniz's basic ideas first appeared in print in 1684, in the *Acta Eruditorum* of Leipzig. The publications of Leibniz were rather short and cryptic, and the first to understand them was Jacob Bernoulli, who in 1687 had become professor of mathematics at Basel. Jacob taught his younger brother Johann some of the secrets of infinitesimal calculus, which he had rediscovered without any help from Leibniz. Thus, by 1690, Newton, Leibniz, and the two Bernoulli brothers were the only people who could handle differential and integral calculus. Apparently, at this time the competition between the two brothers began. Johann's challenge to find the shortest line between two points on a surface was met by Jacob, who in 1698 solved the problem for all surfaces of revolution. The restriction to these surfaces had originally been proposed by Johann himself at the time he stated the problem. Johann approved of his brother's results, but criticized them as too special because Jacob had treated only the surfaces of revolution. In his own paper, Johann announced that he had found the solution of the shortest connection problem for an

David Hilbert (1862–1943) was a leading mathematician of his time.

Frontispiece of the *Acta Eruditorum* 1696.

arbitrary surface. Was Johann's claim really justified? A letter Johann wrote to Leibniz in August 1698 seems to confirm his assertion. In this letter, he stated the "law of the osculating plane." Leibniz quickly replied, praising Johann's solution. About thirty years later, in December 1727, Johann again posed the problem of shortest connection, this time to his student Euler, and this is what led to Euler's 1728 paper on this subject. The appearance of Euler's paper did not keep Johann from attempting to reassert his discovery of the solution to the shortest connection problem. In 1742, he claimed to have communicated the solution by letter to his colleague Klingenstierna in 1728.

Johann's claim, which is accepted today, was long doubted, and one may wonder why. His boasting nature and underdeveloped sense of generosity certainly played a role. The affair connected with the publication of l'Hospital's calculus book also shed doubts on Johann's veracity. This is what happened. In 1691–1692, on the occasion of a tour abroad, Johann met to the Marquis de l'Hospital, the best French mathematician of the time. In exchange for a considerable financial pension, Johann pledged to communicate all the newest discoveries in infinitesimal calculus to the Marquis and to remain silent about their deal during l'Hospital's lifetime. The Marquis transformed Johann's communications into the first textbook on differential calculus, which appeared in 1696. Johann was not pleased but, because of their prior agreement, said nothing. Only after the death of the Marquis and without mentioning his deal with l'Hospital did he claim priority over a part of this book.[9] People then became suspicious, and the reputation of this great but money-loving mathematician certainly suffered. In 1921, Johann's own manuscript on differential calculus was discovered. Together with the correspondence of l'Hospital and Bernoulli, it proved that the Swiss mathematician was the true author of l'Hospital's calculus book.

Even more important than the shortest-connection problem was another question raised by Johann in June 1696. It was to become

[9]The celebrated l'Hospital's rule for finding the value of expressions of the form 0/0.

known as "the problem of quickest descent" or "the brachystochrone problem":

> For two given points A and B in a vertical plane, find a line connecting them on which a movable point M descends under the influence of gravitation from A to B in the quickest possible way.

You may think of a small, heavy ball that slides without friction in the shortest possible time on some track from A to B. What is the shape of this track? It certainly is not the straight line joining A and B. In 1638 Galileo had incorrectly claimed that the solution was a circular arc. What is the correct solution?

It turns out to be a well-known and, by this time, well-studied curve, not a circular arc but a line closely related to the circle: the *cycloid*—that is, the rolling curve of a circle—which we first met in Chapter 2.

The famous French mathematician Blaise Pascal studied the cycloid in 1649, when he was suffering from a painful toothache. To distract himself from the pain, he began to investigate the rolling curve. When the pain disappeared, he took it as a sign that God was not displeased with his thoughts. Pascal's results stimulated other mathematicians to look at this curve, and numerous remarkable properties were subsequently found. The most interesting one was discovered by the Dutch physicist and mathematician Christiaan Huygens.

Huygens was probably most famous for his invention of the pendulum clock, for which he obtained a patent from the Dutch states in 1657. Yet Huygens was not content with his invention, because he

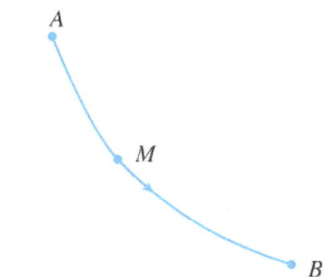
The problem of quickest descent.

Christiaan Huygens (1629–1695).

A cycloid.

A circular pendulum.

The orbit of the tip of a pendulum.

realized that the time a pendulum needed to swing a full period was dependent on its amplitude. In other words, the circular pendulum was not isochronal; it did not always swing to and from in the same amount of time, and this unfortunately led to irregularities in the clocks. So, Huygens asked, is there a *perfect pendulum*? If so, he thought, it should be constructed in such a way that the tip of the pendulum would swing on an *isochrone* (or *tautochrone*). This is a curve on which a frictionless point-mass will always move from any initial point A to the lowest point L of the curve in the same amount of time, regardless of the height of A above L.

Huygens found that the isochronal curve was nothing but a cycloid. Thus, the greatest clockmaker of all times, as he has been called, devised in his 1673 treatise *Horologium Oscillatorium* a pendulum clock whose pendulum would swing on a cycloid. The apparatus is shown below.

The pendulum clock of Huygens. The pendulum swings between two cycloidal cheeks that are shown at the right. From *Horologium Oscillatorium* (1673).

The technical artifice was to use a flexible pendulum that swung between two cycloidal cheeks. This artifice was based on Huygens's beautiful mathematical insight that the *involute* of a cycloid is again a cycloid. This means the following. Imagine a flexible but unstretchable thread laid along an arc E of a cycloid that extends from A to B (see the drawing on the next page). Unwind the thread from E by pulling it tightly so that its loose end always extends tangentially away from curve E. The end point B will describe a curve L that is called the *involute*, whereas E is called the *evolute* (from the Latin word *evolvere*, which means to unwind). Both the involute and the evolute of cycloids, then, are cycloids!

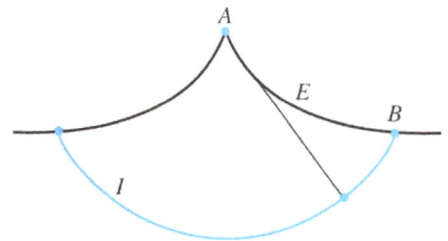

The perfect pendulum: the evolute of a cycloid is a cycloid.

Thus, we see that at the time of Johann Bernoulli's challenge in 1697, the cycloid had been well studied—and so we can imagine how surprised mathematicians were to find that it was the very *brachystochrone* that Johann asked his brother Jacob to find.

In December 1696, Johann published his challenge in the *Acta Eruditorum*, asking for a solution by Easter 1697 and reporting that Leibniz had already found one. In time, five mathematicians solved the problem: Johann and Jacob Bernoulli, Newton, Leibniz, and l'Hospital. Let us see how Johann solved the problem of quickest descent.

The main idea of his proof consisted in transforming the mechanical problem of quickest descent of a heavy point-mass into a problem of optics. Namely, Johann noticed that Galileo's law of a body falling along a certain curve meant that this body would move in the same way as a particle of light would in an atmosphere with a certain density that depended on height above the ground. Johann then used a result of the French mathematician Pierre de Fermat, a judge in the city of Toulouse and one of the greatest mathematicians of all time. Fermat had found that the *law of refraction of light* follows from the assumption that *light always propagates in the quickest way from one point to another*. (This assumption is now called *Fermat's principle*.[10])

[10]Heron's proof of the law of reflection is another application of Fermat's principle.

Pierre de Fermat (1601–1665).

Observing the darkening of Jupiter's moons, the Danish astronomer Ole Rømer could prove that light propagates with finite speed. The figure (an engraving taken from O. Horrebow's *Operum mathematico-physicorum*, vol. 3, Copenhagen, 1741) shows Rømer with his celebrated precision instruments including a pendulum clock, fundamental to his observations, and his new telescope.

The law of refraction (in Fermat's time attributed to René Descartes, but now known to have been discovered by the Dutchman Willebrord Snell (1591–1626) says that the sines of the angles of refraction at a surface separating two homogeneous optical media are inversely proportional to the ratio of their densities, or

$$\sin \alpha_1 / \sin \alpha_2 = n_2 / n_1$$

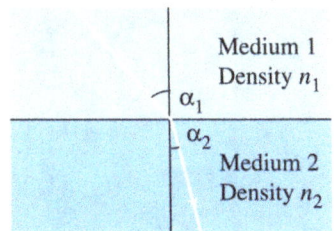

The law of refraction.

The velocity of a particle of light in a medium is inversely proportional to the optical density of this medium.

Johann decomposed the atmosphere into very thin layers of constant density, applied the law of refraction, and then performed a limit procedure by letting the widths of the layers shrink to zero. This led to a differential equation for the curve of quickest descent, and Johann recognized that its solutions were cycloids.

It is remarkable that Johann Bernoulli had successfully connected the fields of optics and mechanics more than one hundred years before the great Irish mathematician William R. Hamilton (1805–1865) was to develop his variational principles, ideas that eventually led to the unification of geometry and physics in present-day science.

DE BEGHINSELEN DER WEEGHCONST

BESCHREVEN DVER

SIMON STEVIN

van Brugghe.

Tot Leyden,
Inde Druckerye van Christoffel Plantijn,
By Françoys van Raphelinghen.
cIɔ. Iɔ. LXXXVI.

Title page of the first edition of Simon Stevin's *De Beghinselen der Weeghconst*, Leyden, 1586.

4 A Miracle and Not a Miracle

Namely, because the shape of the whole universe is most perfect and, in fact, designed by the wisest creator, nothing in all of the world will occur in which no maximum or minimum rule is somehow shining forth.

Leonhard Euler, 1744

In this chapter we want to discuss how equilibrium states of a physical system can be described mathematically. In particular, we would like to know how *stable* equilibria are to be characterized, as, for example, everyone prefers to live in a stable home. It will turn out that these problems can be translated into certain questions related to the calculus of variations. By introducing the concept of the potential energy of a physical system, one can characterize the system's stable equilibrium states as those states that have less potential energy than any other nearby state of the system. This rule describing stable equilibria

by a minimum principle is a special form of Johann Bernoulli's *principle of virtual work*. While the Maupertuis–Leibniz–Euler action principle characterizes true motions among all virtual (i.e., possible) ones, Bernoulli's principle singles out the stable rest positions of a system among all virtual positions. This means the action principle and the principle of virtual work are of similar nature; one governs the nonstationary processes, the other one rules the states at rest.

The principle of virtual work did not simply fall from the heavens, but rather developed over a long period, probably beginning with the fundamental contributions of Archimedes in antiquity and later resumed anew by the Dutch engineer Simon Stevin (1548–1620) in more recent times. Notice the inscription WONDER EN IS GHEEN WONDER, which means "a miracle and not a miracle," in the emblem on the facing page. This emblem appeared on the title page of a work by Stevin and refers to his celebrated solution of an old mechanical problem.

Stevin wanted to know how much pull is needed to keep a weight that lies on a inclined plane in equilibrium (see the illustration at the top of the next page). Clearly, no force is needed if the plane is horizontal, because then it carries the load, and the greatest pull is required when the plane is vertical. Thus these two cases are the extremes, and the general case is somewhere in between: the force necessary to balance the load will depend on the angle of inclination.

What is the precise rule? Stevin found the answer by looking at a slightly more general problem. He considered a prism of height h, the base of which lies on the ground and is supposed to be perfectly horizontal. Assume that the prism has a triangular cross section ABC (see the illustration). The faces of the prism are formed by two

Load on a horizontal plane, on a wall, and on an inclined plane.

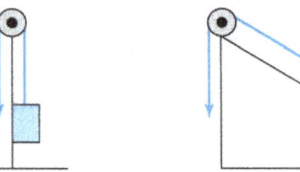

planes, P_1 and P_2, which encloses the angles α_1 and α_2, respectively, with the horizontal ground. Then two loads of weights m_1 and m_2 on P_1 and P_2, respectively, balance each other if and only if $m_1/m_2 = \sin \alpha_2/\sin \alpha_1$. This is the law of the inclined plane, the proof of which appears on Stevin's emblem.

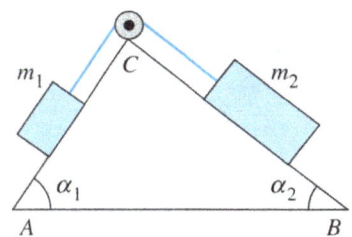

A prism supporting two loads.

Here is Stevin's reasoning. Consider a closed chain of uniform thickness and density that surrounds the prism. Then, Stevin argues, either the chain is in equilibrium or it is moving. If it is moving, then the chain must be in permanent movement, because nothing else happens. Stevin thought this, however, to be impossible. Hence the chain is in equilibrium. The part of the chain below side AB will also be in equilibrium, because it hangs freely without support. Therefore, the upper parts AC and BC of the chain balance each other. Their weights m_1 and m_2 are obviously proportional to the distances \overline{AC} and \overline{BC}; therefore $m_1/m_2 = \overline{AC}/\overline{BC}$. On the other hand, in the rectangular triangle ADC in the figure above, the sine of the angle α_1 is the ratio of the side h to the hypotenuse \overline{AC}, that is,

$$h = \overline{AC} \cdot \sin \alpha_1$$

and similarly we obtain

$$h = \overline{BC} \cdot \sin \alpha_1$$

by looking at the rectangular triangle BDC. These two formulas yield

$$\overline{AC} \cdot \sin \alpha_1 = \overline{BC} \cdot \sin \alpha_2$$

or

$$\overline{AC}/\overline{BC} = \sin \alpha_2/\sin \alpha_1.$$

Combining this formula with $m_1/m_2 = \overline{AC}/\overline{BC}$, we arrive at the desired relation

$$m_1/m_2 = \sin \alpha_2/\sin \alpha_1.$$

If we now replace parts AC and BC of the chain by weights m_1 and m_2, they will balance each other as well.

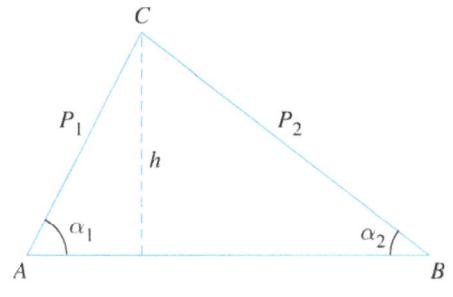

Cross section of a prism.

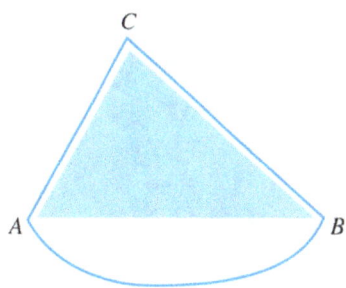

A prism with a chain around it.

Frontispiece of the *editio princeps* of Archimedes by Johannes Herwagen (1544).

A pulley from Ozanam's *Dictionnaire Mathématique* (1691).

Stevin's reasoning is truly remarkable, since it reduces a difficult problem to a nearly obvious statement. The line of thought is quite in the tradition of Archimedes. Stevin was indeed one of the first since antiquity who had read and understood Archimedes's work on statics and hydraulics, and his main work was a modernization of the Archimedean ideas.

Only two of Archimedes's contributions to mechanics were handed down to us. They were called *On Floating Bodies* and *On the Equilibrium of Planes or Centers of Gravity of Plane Figures*. Both were published for the first time in 1543 by Nicolo Tartaglia (1506–1557), an Italian mathematician, just in time for Stevin. The *editio princeps* of Archimedes's complete works in Greek and Latin, edited by *Thomas Gechauff* using the Latin translation by *Jacopo da Cremona*, was produced by the famous printer *Johannes Herwagen* at Basel in 1544.

In his *Equilibrium*, Archimedes left the realm of pure mathematics for that of applied science. He derived a theory of equilibrium for mechanical configurations, based on seven axioms that he placed at the very beginning, thus doing for mechanics what Euclid had done for geometry. In this way he was the first to establish a mathematical theory for a part of physics, an accomplishment that was to become of great importance for the further development of science.

In *Equilibrium*, Archimedes dealt with the lever, which, together with the wedge, inclined plane, roller, and pulley, belonged to the simple machines used in antiquity. (These machines enabled the classical cultures to erect such astounding constructions as the pyramids in Egypt, the Greek temples, and the Roman aqueducts.) He freely used the concept of the *barycenter*, or *center of gravity*, but never gave an explicit definition of this notion. There have been various speculations why Archimedes never defined what he meant by "center of gravity." Some scholars assume that the concept had already been defined, either by earlier scientists or by Archimedes himself in a work now lost. The matter is not decided, but we have to assume

Pulley from the Manesse manuscript (about 1330, University Library Heidelberg).

that the reader was supposed to know what the term "center of gravity" meant. Pappus of Alexandria in about A.D. 340 gave the following explanation (which certainly is correct for convex bodies, though unsatisfactory in general):

> We say that the center of gravity of any body is a point within that body such that, if the body were suspended from the point, the weight carried thereby remains at rest and preserves its original position.

(If the body is convex, that is, if a straight line connecting any two points of the body is also contained in the body, Pappus's definition makes perfect sense because in this case the center of gravity lies within the body. In general, however, the center of gravity may not be contained in the body itself. We then have to modify the above definition in a suitable way. For example, we could consider the convex hull of

A pulley in Huygens's pendulum clock (taken from *Opera Varia*, vol. 1, Leyden, 1724).

Two convex bodies in the plane and their centers of gravity, P.

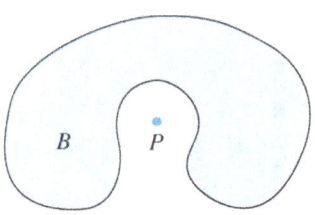

A nonconvex body B and its center of gravity P.

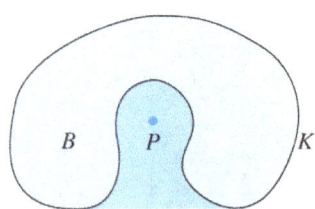

The convex hull K of a nonconvex body B.

the body B which is obtained by supplementing weightless material, rigidly connected with B, in such a way that the resulting body becomes the smallest convex body K containing B. Then the center of gravity of B is simply defined as that of K (see figures).*

To phrase it differently: Knowing the center of gravity of a given body permits one to describe all its *rest positions*. As we shall see shortly, we can even use it to describe all the equilibria of a body and then characterize the stable ones. Archimedes used such insights to describe how forces balance each other and was thereby able to construct very efficient mechanical appliances. Among his most famous insights is his law of the lever, of which Stevin's law is just a variant.

With similar ideas, scientists like Stevin and Galileo started anew, about 1,800 years after Archimedes's death, to build a theory of *statics*, that is, of the equilibrium of complicated mechanical systems. Yet we should not assume that during the intervening centuries mechanical knowledge did not exist. On the contrary, Roman architects developed new concepts, like the arch and the vault, that obviously needed a profound understanding of the principles of statics. Although the arch and the vault had been used by earlier cultures, it was the Romans who perfected them.

One of the admirable buildings of ancient Rome is the Pantheon, built about 1,900 years ago. Originally dedicated as a temple to the seven planetary deities, it later became the Christian church of Santa Maria Rotonda. The main hall of the Pantheon consists of a circular cylinder, covered by a self-supporting cupola of about 140 feet in diameter, constructed of some kind of concrete. The diameter of this cupola was unsurpassed until the year 1890, and even the diameter of the dome of St. Peter's Cathedral in Rome is about four feet smaller.

*Otto Toeplitz (1881-1940) emphasized that Archimedes had given an implicit definition of the barycenter by axioms, just as Euclid had proceeded for geometric notions. This idea was investigated by Toeplitz's student Walter Stein in his paper *Der Begriff des Schwerpunktes bei Archimedes*, Quellen und Studien zur Geschichte der Mathematik, Astronomie und Physik, Abteilung B, Band 1, pp. 221–244 (1931), Berlin, Springer.

The Pantheon in Rome.

The Roman arches can best be admired in the aqueducts found in most countries that were under Roman rule. The purpose of these structures was, and still is, to carry water from the mountains to the cities. The Romans loved water and used it lavishly in fountains, public baths, and private households. Even today, the city of Rome has the greatest abundance of water of all urban settlements in the world, and its wonderful fountains are unique in their exuberance and beauty. It is astounding to

Equilibrium of a rectangular block:
(A) stable position; (B) relatively stable position (that is, stable as long as dislocations are sufficiently small); (C) unstable equilibrium; (D) unstable, with no equilibrium; (E) another stable equilibrium.

think that fourteen aqueducts led to the ancient city, some of them one hundred kilometers long; the few that still function suffice to provide the present city with all the water it needs.

The buildings erected in medieval Europe are possibly even more impressive and astounding than the Roman edifices. With the invention of the pointed arch, based on the ellipse, a new principle of construction was introduced. It permitted construction of the spectacular gothic cathedrals, much admired for their elegance and beauty. At each construction site, a masonry lodge was founded where the craftsmen lived and worked together, often for many decades.

However, we do not know many details of their craft, as each mason had to swear to secrecy. This secretiveness can still be found in the freemasons' lodges that were established in the eighteenth century. Mozart, since 1784 a Mason in the Viennese lodge "Zur Wohltätigkeit," erected a monument to his fellows in his opera *The Magic Flute*, whose solemnity, humor, pathos, and fantastic nonsense have been enjoyed by audiences for over two hundred years. When the opera alludes to the masonic rites, we are reminded of an Egyptian-Pythagorean surrounding. In fact, an early printed text of *The Magic Flute* has an engraving

showing an Egyptian temple, its portal embellished by masonic emblems such as the trowel, the square, and the five-pointed star, which is the Pythagorean pentacle.

During the Renaissance, when the ideas of Archimedes became widely know in Italy, many learned people took up the Archimedean concept of the barycenter of a mechanical system as well as other of his ideas on mechanics. Leonardo da Vinci (1452–1519) had already realized that there is no *perpetual mobile* (perpetual motion machine), and he contrived many engines and mechanical appliances, most of which were never built since the art of engineering was not sufficiently advanced. Guido Ubaldi (Guidobaldo Marchese del Monte, 1545–1607), Galileo, and Torricelli explained and developed the ideas of Archimedes. They stated the principle that such a system will be in (stable) equilibrium if its barycenter is as low as possible within the given limitations. Johann and Jacob Bernoulli, who were to make the city of Basel a world center of mathematics, pondered what might be the equilibrium position of a heavy chain fixed at its end points. They decided that a chain is nothing but a mechanical system consisting of very many small, rigid parts, its links. Hence the equilibrium state should be characterized by the lowest position of the center of gravity. Once the chain has reached a configuration in which it cannot lower any link without in turn raising another, it will be in equilibrium. After having formulated this principle, Johann Bernoulli in 1690 found the exact mathematical shape of the freely hanging chain, as did Huygens and Leibniz. This shape is described by a curve that today is called the *catenary* (i.e., chain line; mathematically, it is a *hyperbolic cosine*). We shall meet the catenary again later, in Chapter 5, which deals with minimal surfaces.

Great progress in the theory of equilibria of mechanical systems was made in 1717 by Johann Bernoulli, who proposed *the principle of virtual work*[1] (which had already appeared in various contexts) as the

Frontispiece of Guidobaldo del Monte's book *Mechanicorum Liber,* Pisa (1577). The figure depicts the Archimedean lever moving the Earth.

The catenary: the equilibrium position of a chain.

[1] This principle was stated in a letter by Johann Bernoulli to the French physicist Pierre Varignon (1654–1722), written January 26, 1717. It was first published in Varignon's *Nouvelle Mecanique*, vol. 2, p. 174, in 1725.

Equilibrium positions of the human body, to be observed by the painter: The barycenter should be supported to keep the equilibrium. (From Leonardo's treatise on painting, *Tractat von der Mahlerey*, Nuremberg, 1747, Second Edition.)

A ball on a hilly landscape.

fundamental law of statics. It states that *in equilibrium, no work is needed to achieve an infinitesimal displacement of a given mechanical system.* This rule captures not only stable configurations, but also unstable ones. To see what this means, imagine a ball of steel that can roll on a landscape with depressions and elevations, as shown above. If the ball is on top of a hill, it lies in an unstable equilibrium; in the center of a pit, it is in a stable position. On the hill, any small kick will move the ball far away from its equilibrium position; in the pit, the ball tries to return to its site of rest, whatever the kicking direction may have been. If, however, the ball lies on a saddle, it is in an unstable equilibrium position. There are directions of no return, yet there are also directions in which the ball may undergo an oscillatory motion after receiving a slight kick. No work is needed (at least in first approximation) to move the ball slightly from any of the three equilibrium positions, because at these points gravity acts perpendicularly to the tangent planes to the surface and is therefore completely balanced by the reacting force of the ground.

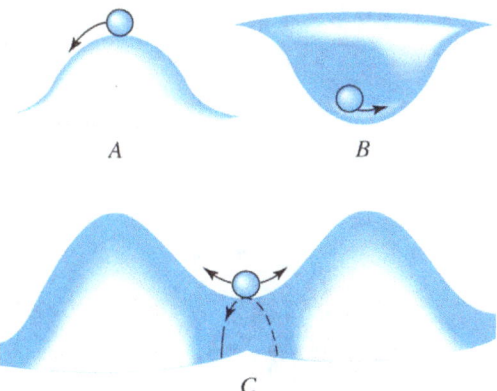

Various equilibrium positions of a ball: (A) unstable; (B) stable; and (C) unstable.

It is not easy to formulate the principle of virtual work in general without employing some mathematics; yet we can recast the principle in such a way that its basic idea can be explained without too many mathematical technicalities. Let us suppose that a number is attached to each state of a physical system; this number is its *potential energy*. We can therefore consider the potential energy to be a function of the various states of the system. Then the principle of virtual work can be expressed by the following two rules, which mathematicians often call *Dirichlet's principle*.

■ **RULE 1.** The stable equilibrium states (that is, the states of rest) of a physical system are characterized by the condition that, in such a state, the potential energy of the system is less than it would be for any possible (or virtual) nearby state of the system.

To get an idea of what this means, we use the geometric picture set up in Chapter 1 when we described minima or maxima of a certain problem as pits or peaks, respectively, of a suitable mountain range. So let us imagine the set of all possible states of a physical system to be a plane, each *point in the plane* representing a possible *state*. If we think of potential energy as the *height* of a mountain range over the plane, then the minima of the range (the pits) correspond to stable equilibria. At these points the tangent plane of the mountain range is horizontal.

As we already know, there are other points that have horizontal tangent planes; these are the maxima (peaks) and the saddle points (passes) of the range. They are supposed to correspond to the *unstable states* of rest. Generally we call the states that correspond to the points of the mountain range with a horizontal tangent plane the *stationary states* of the potential energy. Then we can state the second rule as follows.

■ **RULE 2.** The equilibrium states of a physical system are the stationary states of its potential energy.

This geometric picture yields a simple mathematical framework in which to study equilibria of physical systems. Infinitesimal calculus allows us to define precisely the horizontal tangent planes of the multidimensional mountain ranges that we encounter in physics. Then we can use the methods of the calculus of variations to locate and characterize the equilibrium states. The unstable equilibria are as important as the stable ones. In most physical theories the basic equations characterize the equilibrium states in general only, whereas considerations of stability usually are a difficult matter.

There is, of course, another crucial question: *what potential energy should be assigned to each state of a physical model?* It is here that

Woodcut from Galileo's *Discorsi Dimostrazioni Matematiche*, his foundation of modern mechanics, published by Elsevier in Leyden, 1638.

physical and experimental evidence enter. In order to obtain some feel for the concept of potential energy, let us consider some simple models; but in these models we will not be specific about what the exact potential energy is.

If you are a skier and you gain some height in the mountains either by taking a lift or by walking upward, you also gain potential energy. On top of the mountain you will be in unstable equilibrium. A slight movement starts your race downward. During this race your potential energy is transformed first into kinetic energy and then, by friction, into heat energy. Once you have reached the valley, you will be in a stable position of minimal potential energy. However, if you climb a mountain pass with your skis, you will be at an intermediate unstable equilibrium position, at neither a maximum nor a minimum of the potential energy.

Another example is provided by an elastic medium, say, by the spring of a clock. Such a spring can store a certain amount of energy. As we wind the clock, kinetic energy is transformed into potential energy, which the spring stores and slowly releases as the clock ticks. When the spring has lost its tension, it will be in stable equilibrium and the clock will be at rest.

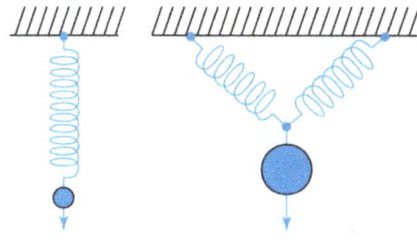

Springs balancing weights.

Similarly, if we squeeze an elastic object like a rubber ball or a rubber ring, we load it with potential energy, which the body would release if it were not for the counterforce exerted by our hands. Once we leave the rubber alone, it will try to lose as much of its potential energy as possible, by reshaping itself to its original position of least potential energy.

If we cut a rubber ring, twist one end, and glue the ends back together, the ring will now contain interior tensions, which it tries to release when it is left free. As a result we will get an equilibrium position that is rather different from the original one.

Still other models can be obtained by loading a system of springs with a heavy weight. The elastic forces of the springs will counteract the weight force exerted by gravity. Once more, equilibrium is achieved if the potential energy of the system is as small as possible.

Rubber cloth carrying a stone or water.

Capillary tubes of varying sizes.

We might replace the springs by a rubber cloth that spans a frame, and place on it some heavy substance, such as a stone or, even more interestingly, some liquid. It is not easy to predict the exact shape of the equilibrium configuration, although it can easily be found by an experiment. Complicating the matter still more, we might add electric or magnetic forces; it will then require a lot of ingenuity and experience to find the correct expression for the potential energy. The field of elastomechanics represents a rather difficult and important topic in applied mathematics, one that is by no means complete despite the important achievements of many great mathematicians and physicists of the eighteenth and nineteenth centuries.

The rubber with the liquid on top is related to a phenomenon we encounter every day: the phenomenon of capillarity. In 1490 Leonardo da Vinci observed the rise and fall of liquids in small tubes, the capillary tubes. This phenomenon lent its name to the theory of surface tension, which is commonly called the *theory of capillarity*. What is capillarity? This is actually hard to describe in a single sentence. Somewhat tautologically, one could say that capillarity phenomena are caused by capillary forces, which are in turn caused by the electric forces by which molecules of a liquid act upon each other as well as upon other molecules, say, those of the container holding the liquid. It makes more sense for us to specify a few phenomena that fall under the heading of capillarity. We have already mentioned the liquids rising in small tubes, studied by Leonardo da Vinci. Other examples for the action of capillarity are water droplets hanging from a surface, as long as they are not too large. One can also notice that the meshes of a sieve retain some amount of liquid. The hairs of a brush cling together if they are wetted. Under leaky faucets drops form and grow to a specific size before falling.

In the common explanation of these phenomena, one imagines that liquid surfaces are covered by an elastic skin that, in addition, may have adhesive or repelling properties if the surfaces are brought into contact with other material. Let us concentrate here only on the elasticity of liquid skins. As a consequence of this property, the liquid skin of a water droplet balances its weight, just as a rubber sheet does so with some liquid on it. This liquid skin explanation of capillarity

Sessile and pendant drops.

phenomena appears to be quite satisfactory, because it is conceptually very simple and can be justified by many experiments. One, in particular, is probably known to everyone through the agility of the water beetle, which can run along the surface of a pond. It walks on the skin of the water.

The British writer Hilaire Belloc described the water beetle's particular talent as follows:

> The WATER BEETLE here shall teach
> A sermon far beyond your reach:
> He flabbergasts the Human Race
> By gliding on the water's face
> With ease, celerity, and grace;
> But if he ever stopped to think
> Of how he did it, he would sink.

The surface tension of water depends very much on the kind and amount of minerals dissolved in it. In some parts of the world, the surface tension of the water is so strong that it can support a small coin. High surface tension makes it much more difficult to wash

The liquid skin of a soap film.

your hands or to clean your laundry. The basic purpose of detergents is to drastically reduce the surface tension of water used in washing machines.

However, it is the reduction of surface tension by soap or by other detergents that enables us to produce soap films, which can either be held by a framework consisting of wires, threads, and surfaces or exist without a boundary, in the form of a soap bubble. Without the soap to reduce the tension, a liquid film of water could not persist, but would immediately break apart.

A precise explanation of surface tension in terms of the action of molecular forces is not easy. Moreover, it must take other phenomena into account—for example, the vapor layer covering a liquid surface. Yet the beautiful and simple concept of characterizing stable equilibrium states as minima of potential energy will enable us to explain the behavior of soap films quite satisfactorily. Because a liquid skin behaves

in many ways like an elastic rubber cloth, it should have higher potential energy the more it is stretched. That is, the stored energy of a liquid surface should be related in some way to its area; the simplest rule is for potential energy to be in direct proportion to area. Thus a soap film will be in stable equilibrium if its area is less than that of any other surface satisfying the same restrictions.

For a liquid drop, the potential energy is more complicated. It depends on the area of the free surface, the gravitational potential of the droplet, and the potential of the forces exerted by the boundary on the liquid. The boundary forces can be either pulling, as between glass and water, or repelling, as between glass and mercury. This approach which is based on Johann Bernoulli's principle of virtual work, was used by Gauss in his 1830 paper "Principia Generalia Theoriae Fluidorum in Statu Equilibrii (General Principles of the Theory of Fluids in Equilibrium). In this work Gauss put the theory of capillarity that had been proposed by Thomas Young ("An essay on the cohesion of fluids," *Philosophical Transactions of the Royal Society*, vol. 95 (1805), pp. 65–87), and also by P. S. Laplace in the supplement to the tenth book of his famous treatise *Mécanique Celeste* (*Celestial Mechanics*, 1806) on a solid basis. The importance of Gauss's work for us is that we now may characterize equilibrium states of a liquid surface by the principle of virtual work, that is, by a variational principle, and so capillarity phenomena can be seen just as another situation where this powerful principle is valid. Soap films are merely a special case of capillarity phenomena, and therefore we can use Bernoulli's principle in the form devised by Gauss to describe the equilibrium positions of soap films as surfaces that, for given boundary configurations, assume a minimum area among all virtual (possible) positions or are at least stationary states of the area. Thus we have arrived at the theory of minimal surfaces; we shall exhibit some of its most interesting features in the next chapter.

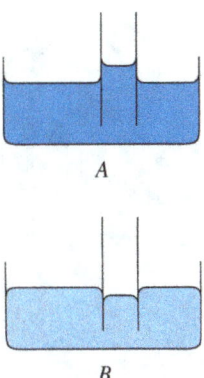

Capillary surfaces: (A) water; (B) mercury. These demonstrate the attracting and repelling forces exerted by the sides of a vessel upon a fluid.

Frontispiece of Gauss's paper from 1830.

A boy blowing a soap bubble. Painting by the French artist Jean Baptiste Siméon Chardin (1698–1779), The Metropolitan Museum of Art, New York.

5 Soap Films: The Amusement of Children and Mathematicians

The book of nature is written in the characters of geometry.

(Galileo Galilei)

Like the boy in the 1739 painting by the French artist Chardin, most of us have at one time or another blown soap bubbles. We prepared a soap solution, dipped a straw into the liquid, withdrew it, and then blew some bubbles.

Bubble blowing is an old pastime. There is even an Etruscan vase in the Louvre, on which children are portrayed blowing bubbles. Today special bubble solutions can be purchased that come with a simple wire or plastic frame and handle. When this frame is dipped into the liquid and then withdrawn, a disklike soap film is formed. Blowing into the frame creates a bubble. If we replace the circular frame with one of a more complicated geometric shape, made of thin flexible wire, beautiful soap films of a bizarre and interesting form will appear. Due to the wave nature of light, the films will soon show iridescent colors produced by interference effects. After some time, however, the film may get so thin that it actually becomes invisible.

SOAP FILMS AND MINIMAL SURFACES: THE PROBLEM OF PLATEAU

The fascinating laminae that appear in soap-film experiments are obviously in stable equilibrium; so, as follows from Johann Bernoulli's principle of virtual work, they must be what are called *laminae of minimal potential energy* that span the wire frame.

Consequently, since potential energy is proportional to area, the *mathematical surfaces* that model soap films are surfaces of least area or, as mathematicians say, minimal surfaces. A minimal surface has a smaller area than any *nearby surface* that spans the same frame. The marvelous shapes of soap films and of their mathematical models are visible and very striking examples of a minimum principle at work.

Mathematicians have long been attracted by minimal surfaces because of the interesting and challenging problems they pose. It was Lagrange, Euler's successor at the court of Frederic II, who in 1760 derived the celebrated *minimal surface equation* that we will state later on.

The extensive experimental and theoretical work on the phenomenon of capillarity by the Belgian physicist J. A. F. Plateau included fascinating experiments on liquid films that stimulated the interest of mathematicians. In 1873 Plateau published a large part of his observations,

Joseph Antoine Ferdinand Plateau (1801–1883).

 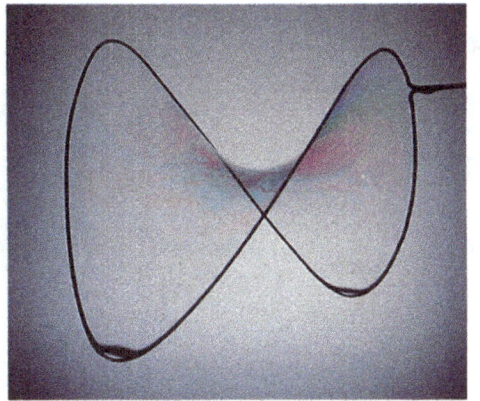

Soap film in a frame.

measurements, and theoretical insights in the treatise *Statique Expérimentale et Théoretique des Liquids Soumis aux Seules Forces Moléculaires* (Experimental and Theoretical Statics of Liquids Subjected Solely to Molecular Forces), his major work.

Plateau did not see many of his experiments; he lost his eyesight in 1843, because during an experiment in physiological optics, he stared at the Sun for longer than 25 seconds without protecting his eyes. In later years he therefore had to rely on the help of his family and assistants when he wanted to carry out his investigations.

One of Plateau's observations was particularly important for mathematics. By numerous experiments, such as dipping various contours into soap solutions and then removing them, he realized that every contour consisting of a single closed wire, whatever its geometric form (provided it is not too large), bounds at least one soap film. Does the corresponding statement hold for the mathematical models of soap films, the mathematical surfaces of least possible area, the minimal surfaces? In other words, is it true that every closed curve in space can be spanned by at least one minimal surface? This *mathematical* question became known as the *Plateau problem*. Many famous mathematicians of the nineteenth and twentieth centuries were intrigued by this problem, but its solution appeared to be very difficult.

Richard Courant (1888–1972) and his student Charles De Prima.

Although the soap film yields an immediate solution to the corresponding physical problem, mathematicians are not spared the effort of proving the existence of solutions to the mathematical problem. This may appear as hair splitting to some readers, but as Richard Courant (1888–1972), one of the prominent researchers on minimal surfaces in this century, has remarked,

> Empirical evidence can never establish mathematical existence—nor can the mathematician's demand for existence be dismissed by the physicist as useless rigor. Only a mathematical existence proof can ensure that the mathematical description of a physical phenomenon is meaningful.

Let us consider an example that illustrates his point. Christian Goldbach (1690–1764) had noticed that even numbers greater than 4 seem to be the sum of two odd prime numbers (it is a mathematical convention that 1 is not a prime number, and 2 is, of course, the only even prime number):

$6 = 3 + 3$;	$16 = 3 + 13$;	$26 = 13 + 13$;
$8 = 5 + 3$;	$18 = 7 + 11$;	$28 = 5 + 23$;
$10 = 5 + 5$;	$20 = 7 + 13$;	$30 = 7 + 23$;
$12 = 7 + 5$;	$22 = 11 + 11$;	$32 = 3 + 29$;
$14 = 7 + 7$;	$24 = 5 + 19$;	$34 = 17 + 17$.

In a letter to Euler in 1742, Goldbach asked whether he could prove that this sequence can be continued *ad infinitum*. Euler was never able to do so, nor has anyone else up to the present day. Today's fastest computer can check the validity of Goldbach's conjecture for an enormously large number of even integers but obviously cannot verify it for all even integers, since there are infinitely many. Thus any checking procedure would require an infinite amount of time if all even numbers were to be tested.

This consideration provides a good reason why physical experiments can never prove a mathematical statement. No matter how many experiments we carry out in order to verify a claim, we can never exclude the possibility that another repetition of the experiment might yield a different result. Hence empirical evidence can never establish a mathematical fact. Even for the physicist it is important to obtain a rigorous mathematical existence proof, because only in this way can it be verified beyond any doubt that a mathematical model of a physical situation is at least in some respect correct. Passing to a mathematical model of soap films necessitates that we speak of a curve instead of a wire and a surface instead of a soap film. Curves and surfaces are idealizations like those created by the ancient Greeks. We have indicated this process in Chapter 3.

The preceding reasoning for the Goldbach conjecture is equally valid for the Plateau problem. The physical problem always appears to

A B C

A closed curve that can be spanned by at least three different minimal surfaces.

have a solution, since every time we dip a contour in soapy water and then withdraw it, a soap film forms. However, from this physical evidence we cannot conclude that every closed contour is spanned by a surface of least area.

The solutions of the Plateau problem can be of many different types. In order to get an idea of what may happen, let us consider a particular closed curve that can be spanned by at least three different minimal surfaces. In the figure at the left, the middle and right-hand surfaces are of the same type, but the left-hand one is of a different kind. Mathematicians say that the left-hand surface has *genus* 1, whereas the other two have *genus* 0. What does this mean?

To solve the Plateau problem, we need to know how complicated a surface-spanning contour can be. We thus need a way to classify the surfaces among which we might find a solution. A coarse but useful classification distinguishes between various *topological types*. Two surfaces are of the same topological type if we can deform one into the other by stretching as if it were an elastic membrane, allowing different parts of a surface to pass through one another (for example, see the illustration on page ???), but without any tearing, cutting, or gluing. Some surfaces of the same topological type as a sphere are shown below.

Closed surfaces of the same topological type as the sphere.

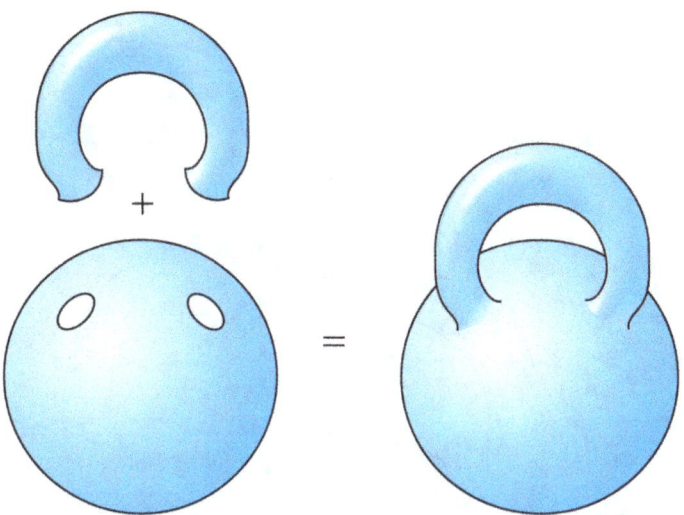

One handle being attached to a sphere.

We may now cut out two disks from the sphere and attach a handle (see the figure above). We can continue this process, attaching two, three, or more handles to the sphere. In this way we get a collection of finite surfaces without boundary, called *handle–body surfaces*, that are mutually distinct. If two such surfaces have a different number of attached handles, they are of different topological types (see the figure below).

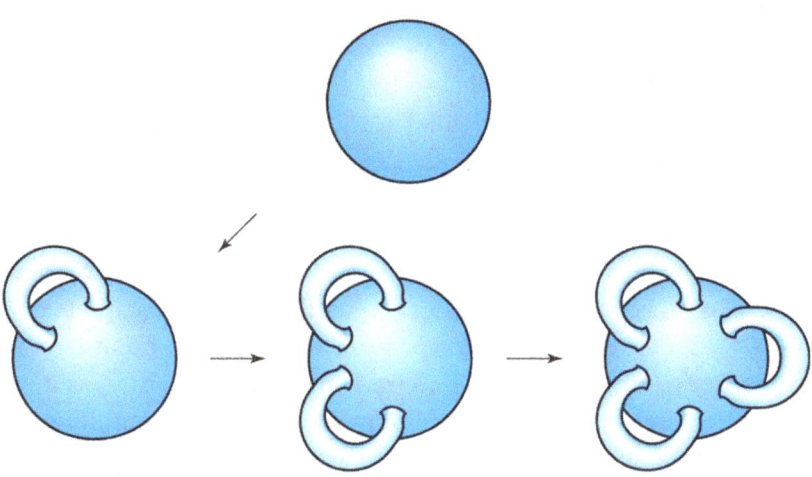

A sphere with 0, 1, 2, 3 handles attached.

Surfaces of genus 1 (doughnut), genus 2 (pretzel), and the baker's pretzel.

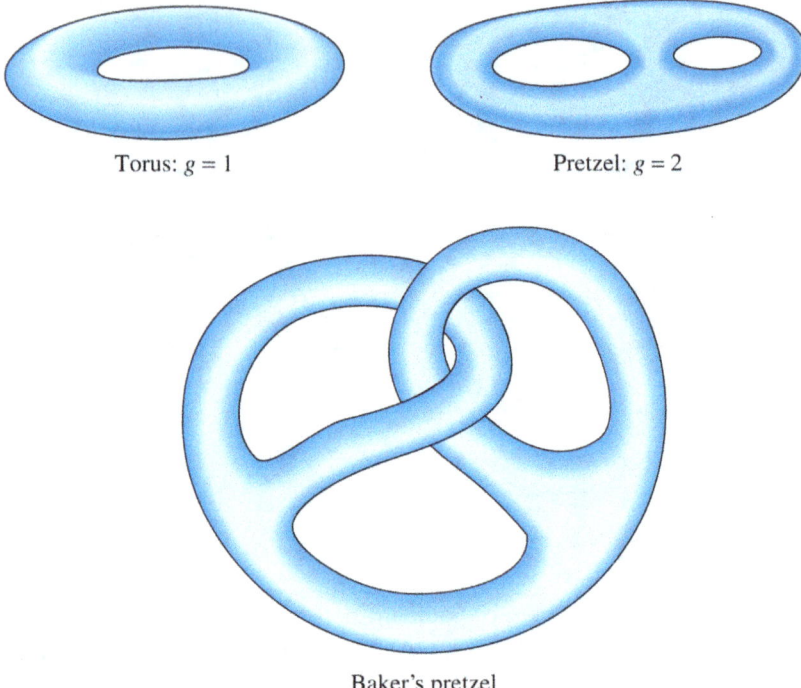

Torus: $g = 1$

Pretzel: $g = 2$

Baker's pretzel

Observe that stretching the sphere with one handle turns it into a torus, the surface of a doughnut; also, the sphere with two handles is of the same type as the surface of a pretzel. Notice that the baker's pretzel shown above is of the same topological type as the sphere with two handles.

A surface is said to be of *genus g* if it is of the topological type of a sphere with g handles attached. Thus the torus has genus 1, and the pretzel has genus 2.

The surfaces that we have considered so far have not had any boundary; that is, when we walk around the surface we never come to an edge. For example, on a sphere, torus, or pretzel we never meet an edge, so they are surfaces without a boundary. On the other hand, the inside of a circle in the plane, a disk, is a surface that has this circle as its boundary. Another example is furnished by a sphere, or a torus into which we cut one or several

holes by removing little disk-like surfaces. Now we may cut holes into them, say, m holes that are bounded by m curves C_1, C_2, \ldots, C_m. Two surfaces obtained in this way turn out to be of the same topological type if they have the same genus and the same number of holes. More specifically, we say that a surface with m holes in its skin has the genus g if, when we fill the holes with m disklike surfaces, we create a surface without boundary and of the type of a sphere with g handles attached. We may call this object a handle–body surface of genus g with m holes.

Let us now consider surfaces with exactly one hole. All of them could, in principle, be of the same topological type as the solutions to Plateau's problem. Indeed, it is quite possible that the same curve may bound minimal surfaces of different topological types. For example, surfaces B and C on page 150 above have $g = 0$ and $m = 1$, whereas surface A has $g = 1$ and $m = 1$.

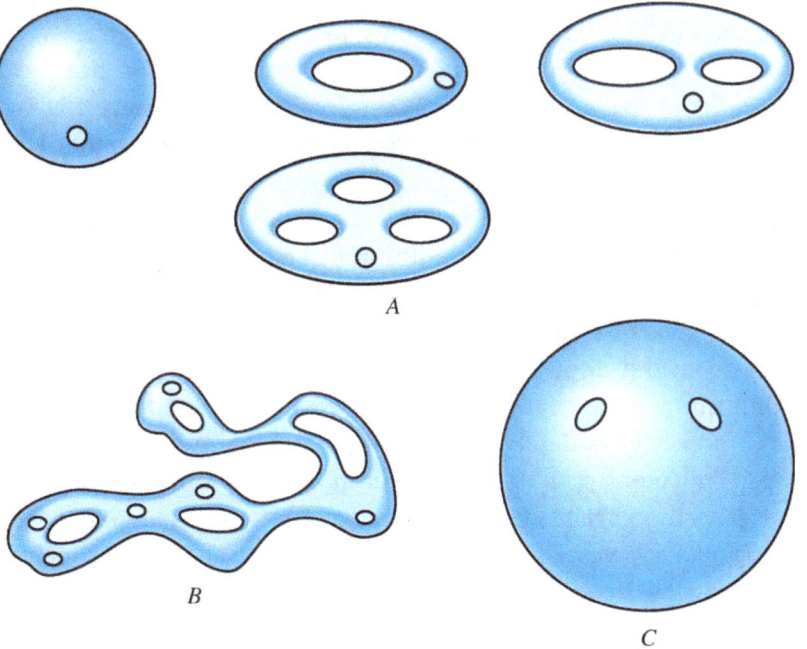

Surfaces with (A) one hole but different genus; (B) surface of genus 4 with 6 holes; (C) sphere with two holes.

A sphere with one hole can be stretched into a disk.

We note that any surface of the type of a sphere with one hole (that is, $g = 0$, $m = 1$) is of the same type as the disk (see the figure). The disk is the simplest possible surface whose boundary is formed by a simple closed curve. Hence all the disk-type minimal surfaces within a given contour are in a way the simplest solutions to Plateau's problem.

A disk-type minimal surface may have an area smaller than that of any nearby surface or, in fact, any other disk surface; yet there may be a surface of higher genus with less area. In other words, a disktype minimal surface may not be the surface of least area that spans a pre-assigned curve but may furnish only a local minimum of area. An example of this phenomenon is displayed in the margin on page 150. (In order to make it very convincing, we should squeeze the two pairs of adjacent circlelike parts of the curve very close together.)

Thus for a given curve C it is not at all obvious which genus g will yield the surface of absolute minimum area spanning C. It is not even obvious if a minimum is achieved for any surface no matter what its genus is. For example, consider a "pathological curve" like the one in the upper illustration on the next page. The absolute minimum area is achieved by a surface that looks like the "Loch Ness monster," as depicted in the figure in the middle of the next page. This monster is not contained in our previous collection, because its topological type is that of *a sphere with only one hole to which infinitely many handles are attached*. Therefore it is topologically a disk with infinitely many handles, as shown in the lower figure on the next page.

The boundary of the Loch Ness monster.

The Loch Ness monster.

A disk with infinitely many handles.

This absurd example arises because we considered a contour that possesses a *singular point:* the very tip of the tail of the monster. Could it be that such unpleasant *surfaces of infinite genus cannot appear as solutions to the Plateau problem if we admit only "regular" smooth curves as possible contours*? This, in fact, has recently been proved.

Do minimal surfaces have other surprises waiting for us? Can we, for instance, expect to find all possible soap films that span a given contour in our collection of handle–body surfaces, or must we extend this list?

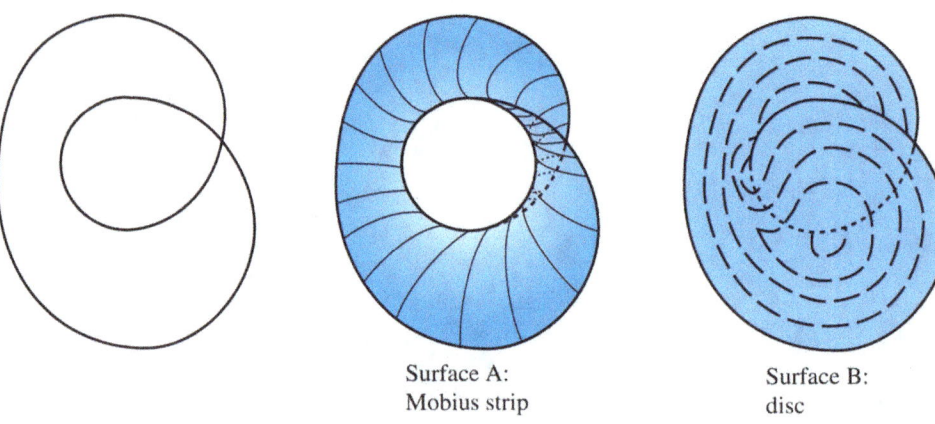

Surface A: Möbius strip

Surface B: disc

Two minimal surfaces spanning a closed curve.

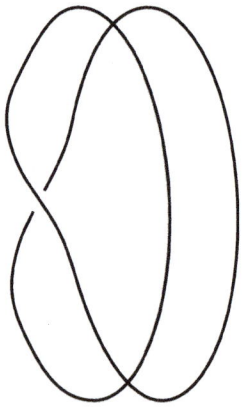

The surface of least area spanning this curve is of the topological type of the Möbius strip.

Let us consider the curve shown above. It is spanned by two surfaces *A* and *B* of different topological types; and, if the wire looks like that in the figure in the margin at the lower left, then a surface very similar to surface *A* will be the surface of least area that spans it. However, surface *A* is not contained in our collection of handle–body surfaces with *m* holes and with genus *g*, because all of those surfaces are two-sided but *A* is one-sided. "Two-sided" means that the surface really has two sides, like surface *B*, so that a small animal walking on the surface can never move from one side to the other without crossing the boundary, but on a one-sided surface it can get anywhere without ever crossing the boundary.

Surface *A* is of the same topological type as the celebrated *Möbius strip* discovered in 1858 by the German mathematician and astronomer A. F. Möbius, professor at Leipzig University. We can produce this strip by cutting a long strip of paper, twisting it once, and pasting its ends together. This surface indeed has only one side, as we can see either by producing it or by studying the 1968 etching by the Dutch artist M. C. Escher, as shown opposite.

Möbius strip (by M. C. Escher).

August Ferdinand Möbius (1790–1868)

Generating a Möbius strip.

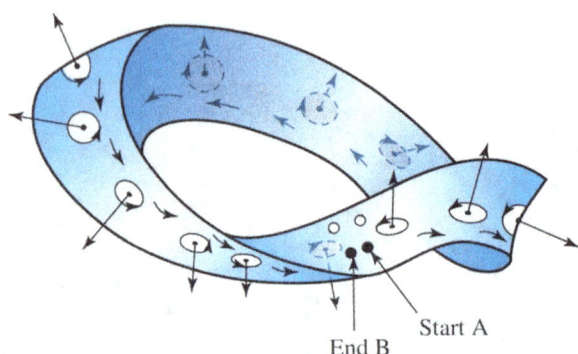

One complete circuit of the Möbius strip changes a clockwise rotation to a counterclockwise one.

The Klein bottle has only one side: its inside is the same as its outside.

A nonorientable surface of genus 1 spanning a closed curve.

A nonorientable surface of genus 2 spanning a closed curve.

Since the discovery of the Möbius strip, mathematicians have found a large set of different topological types of finite one-sided surfaces, with no hole, one hole, two holes, and so forth. However, many of them must have self-intersections, like the Klein bottle, named after the German mathematician Felix Klein (1849–1925). These surfaces are called *nonorientable* because we cannot distinguish a clockwise from a counterclockwise rotation on such a surface (see above and at left here). Our collection of handle–body surfaces represents the set of orientable finite surfaces.

Finally, let us consider two one-sided minimal surfaces that are of more complicated topological types than the Möbius strip. The figures at the left show that it is far from obvious what the topological type of a potential solution to the Plateau problem should be. However, we have already seen that surfaces of the disk type are among the simplest surfaces that can span a closed curve. Thus we pose the following question:

Can we prove that there exists at least one surface of the disk type having a minimum area within the class of all disk-type surfaces that span a given simple closed curve? (A *simple* curve is one that does not intersect itself, like a circle but unlike a figure eight.)

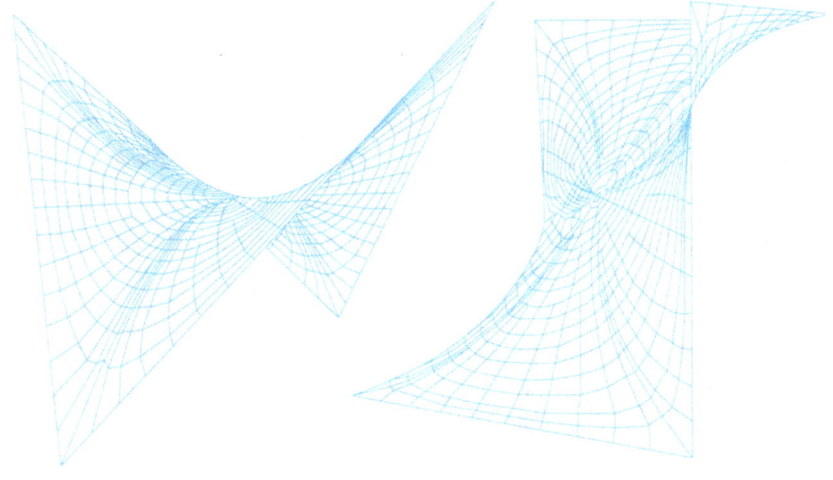

Computer-generated drawings of two of Schwarz's solutions to Plateau's problem for polygons.

During the nineteenth century, this disk version of the Plateau problem was solved for many special curves, most of which were polygons, and many special minimal surfaces were discovered. The achievements of mathematicians like Riemann, Weierstrass, and Schwarz were of first rank; nevertheless, the tools available to them did not suffice to tackle the Plateau problem in general. In 1914, the French geometer Gaston Darboux (1842–1917) observed,

> Until now, mathematical analysis has not been able to invent any method that will permit us to begin with the investigation of this beautiful question.

In 1928, a young American mathematician, Jesse Douglas, solved the disk version of the Plateau problem. In the same year, he lectured about his solution at the University of Göttingen in Germany, at that time one of the principal scientific and mathematical centers of the world. Apparently not all of Douglas's arguments were convincing, and the audience had doubts about the completeness of his proof. It took Douglas two more years to fill in the details, and his work did not appear until 1931. By that time a totally different proof devised by the Hungarian mathematician Tibor Radó had been published.

Jesse Douglas (1897–1965)

Tibor Radó (1895–1965)

During the following decade, Douglas solved many other problems for minimal surfaces. His powerful methods permitted him to prove existence, under certain conditions, of minimal surfaces of higher genus spanning one or even finitely many closed curves. For his accomplishments he was awarded, in 1936, one of the first two Fields medals in mathematics. (Nobel prizes are not awarded in mathematics, and the Fields medal is the highest award offered by the International Congress of Mathematics.)

The work of Douglas and Radó did not end the study of minimal surfaces. Since then, as we shall see, many other beautiful results have been obtained. One of the most important achievements has been the solution of the general Plateau problem. It has now been proved that, *for each simple curve, there exists a surface of least area and of finite genus spanning this contour; furthermore, this least-area surface will not have self-intersections.* From this we can infer that for closed "knotted" curves the least-area surface cannot be a disk; for if there were one, we could prove that the knot could be unknotted. This, however, is impossible for a *closed* knotted curve, and so the only way out of this dilemma is that the curve was not really knotted in the first place.

A GEOMETRIC DESCRIPTION OF MINIMAL SURFACES

In the preceding section, we used the term "minimal surfaces" for the minima of area—that is, for the mathematical models of stable soap films. For such soap films, the potential energy of the "surface tension" is at a local minimum.

We have seen that a given closed curve may bound more than one "stable" minimal surface. It is then natural to ask if there exist surfaces spanning a given contour that correspond to *unstable equilibria of potential energy*. In the mathematical mountain range generated by potential energy, that is, by area, these minimal surfaces correspond to the pits of the range (such as points P_1 and P_2 in the figure on the next page). But in this range we also have another point P_3, where

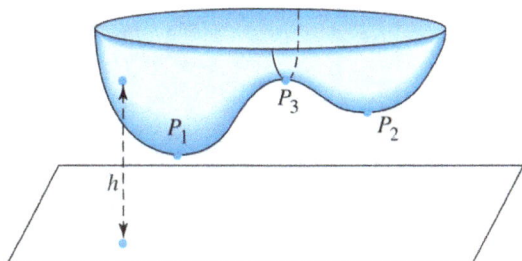

Deformed bowl

there is a horizontal tangent plane but which is not a local minimum. As we have discussed in Chapter 4, such points are also called equilibrium points, but they are points of unstable equilibrium, while minima such as P_1 and P_2 are points in a stable equilibrium.

Another possibility to illustrate the situation would be to use a mechanical analogue: a deformed steel bowl, as we now interpret the figure above. Here we have two minimum points P_1 and P_2 for the height h above a fixed plane, we have a point P_3 of unstable equilibrium, and a little ball put into positions P_1, P_2, or P_3 is in rest. If we give it a little push, it rolls away from the unstable position P_3 while it oscillates around the stable position P_1 or P_2, respectively.

Consider now a closed contour that can be spanned by two different disk surfaces that are local minima of area. We then can ask whether there is a third surface yielding an unstable equilibrium of an area that would correspond to point P_3, the unstable equilibrium. If so, can we find a *geometric condition* that describes both stable and unstable equilibria of an area? In both cases, the answer is "yes." The existence of two minima of an area indeed require the existence of a disk surface that is an unstable equilibrium of that area. We cannot see the corresponding soap film, because of its instability, it is a fleeting ghost that may appear for only a fraction of a second when we flip one minimum into the other. A picture of such a phantom soap film is shown in part (C) of the figure at the top of the next page.

How can we formulate a condition that will characterize all equilibrium surfaces, stable or unstable, in the same way? We begin our quest with a

 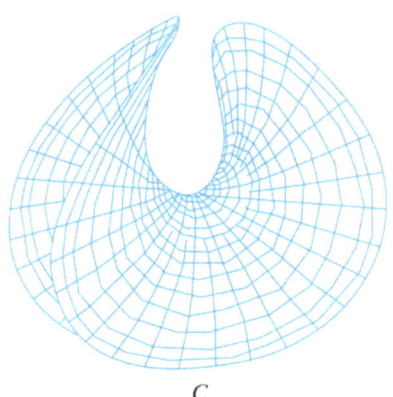

A　　　　　　　　　　B　　　　　　　　　　C

Computer-generated drawings of (A and B) stable minimal surfaces; and (C) of "Enneper's surface," for this choice of the boundary curve an unstable equilibrium surface, which represents a phantom soap film.

discussion of the *curviness* of curves. Everyone knows that a straight line does not curve; indeed, this is a tautology. If we were asked to assign a value to the amount of curviness in a straight line, the answer would, of course, be zero. But what about other curves; how curvy are they?

In order to answer this question we first consider a circle of radius r. A circle curves the same amount at every point. If r is small, then the circle is very curvy. As the radius r increases, the circle becomes larger and larger and curves less and less. We would like a definition of curvature that reflects this simple observation. We therefore *define* the curvature κ (a Greek kappa) of a circle by the following rule:

$$\kappa = \frac{1}{r} \quad \text{or} \quad curvature = \frac{1}{radius}.$$

Thus as the radius r gets larger, the curvature κ gets smaller.

Then, what do we do with a more complicated plane curve, like the one below on the opposite page (part A)? It is obvious that this curve is very curvy in some places and does not curve at all in others. If a number can be assigned to measure the curviness, this number will therefore change as we move from point to point on the curve. Let us take an arbitrary curve C, planar or not, and a point P on it (as shown in part B at the bottom of the next page).

If the curve is smooth, there will be a tangent line T to the curve at point P. We can then show that there is a circle that passes through point P, that has the same tangent as that of curve C, and that "best fits" C at point P. The precise definition of this circle requires infinitesimal calculus, but we can give a rough idea of it. Choose two points P_1 and P_2 close to P. Then the three points P_1, P_2, and P in general determine a circle. As P_1 and P_2 squeeze in toward P, this circle gets closer and closer to the best-fitting one. If r is the radius of this circle, then the curvature κ of C at P will be defined as $1/r$. For reasons that will soon become clear, we shall also give either a positive or negative sign to the curvature of certain curves that are called *normal sections of a surface*. That is, we shall define the curvature κ of a "normal section" as $1/r$ or $-1/r$, according to a convention that we shall now describe.

Circles of increasing radius.

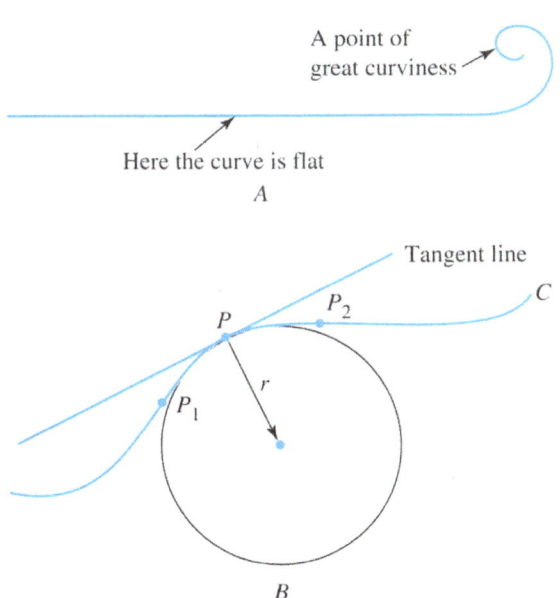

A. A general curve.
B. Radius of curvature.

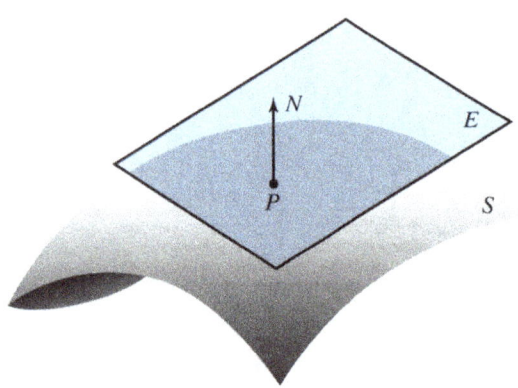

A surface S, a tangent plane E, a point P, and a normal direction N.

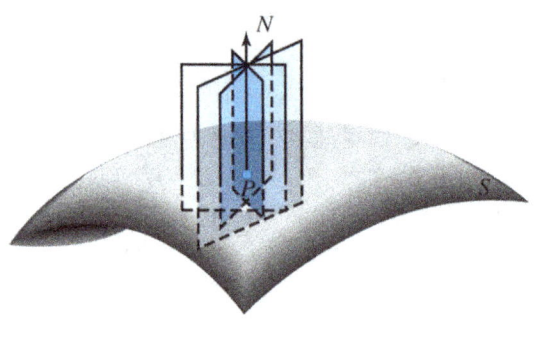

A bushel of normal planes at P determined by a normal direction N.

First, what is a normal section? Let S be an arbitrary surface as shown above at the left. We fix a point P on this surface, and then we choose a direction N that is perpendicular to the tangent plane E that touches the surface S only at the point P. Such a perpendicular direction is called a *surface normal* of S at the point P. One may imagine such a normal N as an arrow that is fixed with its tail at the point P (at each point of a regular surface there exist two normal arrows, pointing in opposite directions).

Two normal planes and their normal sections.

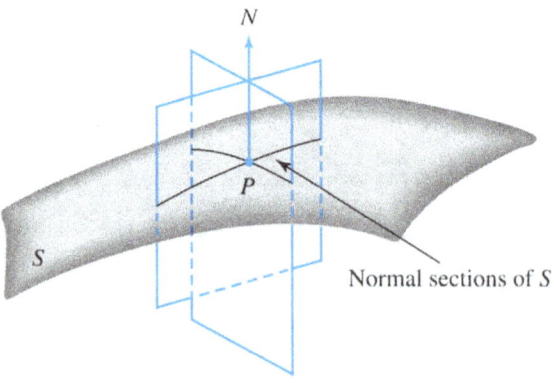

A plane that passes through point P and contains the normal direction N, as shown at the right on the top of the previous page, is called a *normal plane* to the surface at point P. Obviously, there can be infinitely many normal planes at P; we can find them all by revolving any such plane about the axis determined by the normal direction N.

Each normal plane at point P intersects the surface S in a plane curve that is called a *normal section of S*. Consider now a normal section C of the surface at point P as shown in the figure at the bottom of the previous page, and let r be the radius of the best-fitting circle at P. This circle lies in the same plane as the normal section C, and its center is situated on the straight line that contains the "normal arrow" N. Then either the arrow N will point from P to the center of the best-fitting circle or N will point in the opposite direction. If it points into the circle, we define the curvature κ of the normal section at P by $1/r$; if it

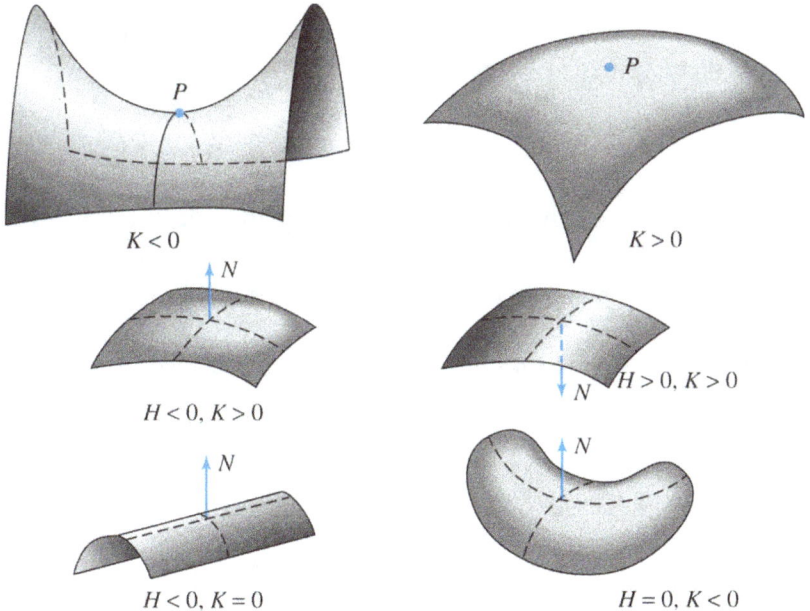

Surfaces with different curvatures. The sign of K has a geometric meaning, whereas the sign of H merely depends on the choice of N, but not on the shape of the surface.

points away, by $-1/r$. Thus κ can be positive or negative depending on the choice of the normal N. Negative κ means that the normal section is curving away from N, and positive κ means that it is curving toward N. Consequently, this sign convention not only allows us to describe the amount of curviness of a normal section at the point P, but includes some directional information as well.

We are now ready to define a measure of curvature for a surface, our next major step. Two such measures of curvature are used by geometers today: the Gaussian curvature and the mean curvature.

Among all normal sections passing through a fixed point P of the surface, one can prove that there exist two special sections, C_1 and C_2, with curvatures κ_1 and κ_2, respectively, such that κ_1 is the largest and κ_2 the smallest curvature that a normal section at point P can have.

We now define the *Gaussian curvature* K of the surface at P as the product of the two principal curvatures κ_1 and κ_2,

$$K = \kappa_1 \cdot \kappa_2;$$

and we define the *mean curvature* H as the arithmetic average of the principal curvatures,

$$H = \frac{\kappa_1 + \kappa_2}{2}.$$

Now we can discuss the theorem of Lagrange in which he stated the minimal-surface equation.[1] This theorem will provide the geometric characterization of least-area surfaces that we are seeking:

> At each regular point, a surface of minimal area must have a mean curvature of zero.

[1] The theorem as presented here was extended by Meusnier, one of the early differential geometers, in his *Mémoire sur la Courbure des Surfaces*, read to the French Academy in 1776 and published in 1785.

That is, the surfaces of minimal area satisfy the equation

$$H = 0.$$

(In the neighborhood of a regular point, a surface is smooth and looks like a slightly deformed disk.)

There is another physical interpretation of this equation. We can relate the pressure difference p between the sides of a soap film at any point to the mean curvature H of the film by *Laplace's equation*

$$p = TH,$$

where T denotes the surface tension of the liquid. The same equation relates the mean curvature of an interfacing surface between two liquids (or a liquid and a gas) to the difference in pressure between them.

Saying that the mean curvature must be zero is, therefore, equivalent to saying that on both sides of a minimal surface we have the same pressure. (In the interior of a soap bubble, however, we find higher pressure than outside of the bubble; thus a bubble is a *surface of constant mean curvature H* different from zero.)

It turns out that the surfaces that furnish an unstable equilibrium of area must likewise have a mean curvature of zero. For this reason it has become customary to use the term "minimal surface" for all surfaces having a mean curvature of zero, whether they minimize area or not. These are the objects that we shall now investigate.

From now on we shall refer to surfaces that are minimal for area (that is, for potential energy) as *stable minimal surfaces*, whereas the term *minimal surfaces* will indicate surfaces of zero mean curvature; such surfaces correspond to states of a soap film that are stationary with respect to the potential energy of surface tension.

Stable minimal surfaces are those surfaces that, in principle, can be realized as soap films. However, the unstable minimal surfaces—the

fleeting ghosts—are also interesting and important, even though the corresponding soap films cannot be produced. First, they are of interest to the mathematician as mathematical objects defined by a beautiful geometric condition. Second, they are needed to help us to understand how minimal surfaces behave as we perturb their boundary contour.

What do minimal surfaces look like? If a surface has zero mean curvature at each point, then at each point it is curving both away and toward a given perpendicular direction. That is, *minimal surfaces either are flat or look like saddle surfaces.*

HOW MANY MINIMAL SURFACES CAN SPAN A FIXED CONTOUR?

A natural question that we can ask about a given contour is *how many minimal surfaces can it bound*? Unfortunately, nearly nothing is known, even if we restrict ourselves to the much simpler question of *how many disk-type minimal surfaces can span a closed curve*? The figure in the margin on page 150 shows that there are certainly contours that can be spanned by more than one disk-type minimal surface, but are there contours that bound only one such surface?

A first answer has long been known: planar curves bound only one minimal surface, a planar one. A second result was found by Tibor Radó in 1932. He showed that, if a contour has a simple projection onto a convex curve in a plane, then it can bound only one disk-type minimal surface.

A curve with a convex simple projection onto a plane. A projection is simpe if each of its points corresponds to only one point on the curve.

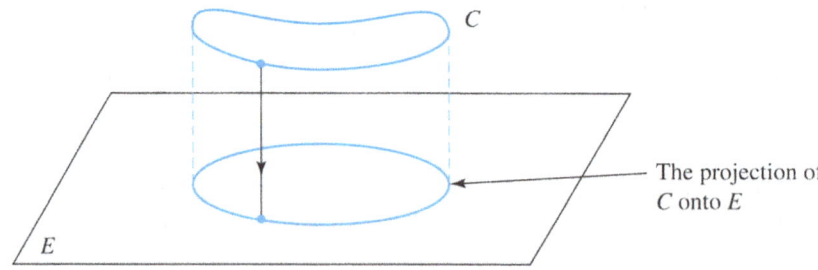

The projection of C onto E

The third and only other mathematical result on uniqueness uses the notion of *total curvature of a spatial curve*. This is a number that measures how much the entire contour curves. For example, a circle with radius 1 has a total curvature of 2π. If only a small part of a contour is very curvy, then the total curvature will not be too large. The following result is known:

> If a given contour has a total curvature of less than 4π, then there can be only one disk-type minimal surface bounded by that contour.

Thus one might hope to find at least an upper bound on the number of disk-type minimal surfaces that span a "not too complicated" contour. This hope has been completely dashed, since it was shown that, given any number N (say 100, 5000, 10^{100}, . . .) and any small positive number ε (say, 1/10, 1/8000, 10^{-200}, . . .), there exists a contour of total curvature less than $4\pi + \varepsilon$ that bounds at least N disk-type minimal surfaces. This striking theorem shows that if the total curvature of a curve is even slightly larger than 4π, then very wild and quite unimaginable things can occur.

Except for the three uniqueness results that we just stated, nothing is known about the number of minimal surfaces or even about the number of disk-type minimal surfaces that span a given nonplanar contour. There is no example of a curve that bounds more than one minimal surface and for which *all* minimal surfaces that span it are known, nor are there even any examples that indicate plausible answers. However, for genus zero, there is an example of a contour that bounds exactly three disk-type minimal surfaces.

The figure at the top of the following page shows two minimal surfaces spanning the same uncomplicated contour; yet it is not known how many more minimal surfaces are bounded by this contour.

The interesting question, as to whether there are contours that bound infinitely many disk-type minimal surfaces, appears now to have been answered. Intuition seems to tell us that a fixed contour will bound only finitely many such surfaces, but intuition can often be misleading in mathematics! There is, in fact, an intriguing curve

Two surfaces spanning the same contour.

that plausibly bounds infinitely many disk-type minimal surfaces, the monster curve depicted on page 155.

Let us see how this curve is constructed. First, take the contour in the top figure on the next page. This curve bounds at least two stable disk-type minimal surfaces (see the figure in the margin on page 150). Next, consider an infinite sequence C_1, C_2, C_3, \ldots of curves of the same kind, selected in such a way that C_2 is half the size of C_1, curve C_3 is half the size of C_2, and so on (see the second figure on the opposite page).

Join curves C_1 and C_2 by a small bridge, (as shown in the third figure on the next page), cutting off some part of each curve to ensure that a simple closed curve C^* is formed. Then we can join each of the two stable minimal surfaces spanning curve C_1 to each of the two stable minimal surfaces spanning C_2 by adding a small ribbon spanning the bridge; so there are at least $2 = 2 = 4$ combinations of disk-type surfaces bounded by C^*. Now it is plausible that if the bridge is small enough, then, very close to each of the four "joint surfaces" generated by the bridging procedure, there exists a stable minimal surface.

A simple closed curve that bounds at least two surfaces.

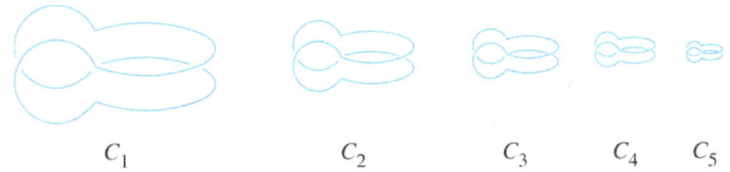

An infinite sequence C_1, C_2, C_3, \ldots of contours of decreasing size (C_{i+1} is half the size of C_i) and tending to a point.

Two curves connected by a bridge.

Four surfaces that span the composite curve C^* in the illustration above.

In this way we have obtained from curves C_1 and C_2 a new simple closed curve that bounds four different stable minimal surfaces of the disk type. Why stop here? Attach to the contour C^* the third curve, C_3, in our sequence by a second bridge half the size of the previous one. We then obtain another simple contour that, by the same plausibility argument as before, will bound at least $2 \times 2 \times 2 = 8$ stable minimal surfaces of the disk type. Next we attach C_4, C_5, C_6, and so on, by bridges of increasingly smaller size. By repeating this process indefinitely, we arrive at the simple closed curve C of finite length pictured on page 155. At each stage of the connecting procedure we obtain at least twice as many minimal surfaces within the new contour as we had counted within the previous one. Thus we can argue that C bounds infinitely many disk-type minimal surfaces, and this, in fact, has been proved rigorously.

On the other hand, the situation is not quite as bad as it may now seem. In fact, it has been proved in recent years that if *one picks a simple smooth contour at random, the probability is zero that it bounds infinitely many disk-type minimal surfaces*. In other words, the "bad curves" are extremely rare; almost all smooth curves bound only a finite number of minimal surfaces of the type of the disk. This result does not contradict the statement that the monster curve bounds infinitely many minimal surfaces since it might belong to the set of bad curves. Moreover, the monster curve is not a smooth curve, even if we attached all bridges in a smooth way. The very tip of the tail of the monster curve is an extremely irregular point, a kind of kink.

For "very smooth" curves (the technical term is *analytic curves*) something more is known. Such a curve can bound only a finite number of least-area surfaces among all surfaces of the type of the disk.

THE ALGEBRAIC FORMULA RELATING THE MINIMAL SURFACES THAT SPAN A FIXED CONTOUR

Despite all the peculiarities discussed in the preceding section, a method has recently been discovered to relate *all* possible disk-type minimal surfaces bounded by a fixed wire. To explain this result, let

us return to our simple model of the deformed steel bowl situated over a plane as shown in the figure on page 161. If, for this bowl, we consider the height h above the plane, then there are three equilibrium points for h, the two minima P_1 and P_2 and one unstable equilibrium point P_3. Imagine a spider walking along the inside of this bowl. If the spider starts at either P_1 or P_2, then its height above the plane will increase regardless of the direction in which it walks. However, at P_3 the situation is different. There the spider can choose a direction along which its height above the plane decreases. Like our skier in Chapter 4, the spider is decreasing its potential energy. Let us give a name to the maximal number of mutually perpendicular directions in which the spider can walk away from the equilibrium point and yet decrease its height; we will call it the *characteristic number* of the equilibrium point. For P_1 and P_2 the characteristic number is 0; for P_3 it is 1.

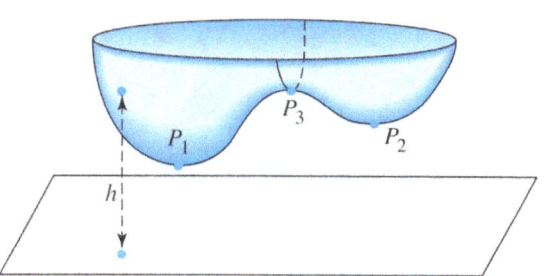

A deformed bowl.

The figure at the right shows an upside-down bowl over a plane E. The only equilibrium point is P, the top of the inverted bowl. Whatever direction the spider crawls from P reduces its height; so the characteristic number of P is 2.

A spider on an upside-down bowl.

Let us return to our first bowl (see the figure above), where we have the three equilibrium points with characteristic numbers $\lambda_1 = 0, \lambda_2 = 0, \lambda_3 = 1$. We then form the sum

$$(-1)^{\lambda_1} + (-1)^{\lambda_2} + (-1)^{\lambda_3} = 1 + 1 - 1 = 1.$$

Now suppose we take *any* hemispherical bowl and deform it (but not along its border), creating in this way some number of pits, passes, or peaks, say, n equilibrium points P_1, \ldots, P_n with characteristic numbers $\lambda_1, \ldots, \lambda_n$ (see the figure at the right).

The reader may check that all these examples will satisfy the remarkable equation

A valley with three pits.

$$(-1)^{\lambda_1} + (-1)^{\lambda_2} + \cdots + (-1)^{\lambda_n} = 1.$$

This equation, called *Morse's equation*, is forced upon us by some mathematical structure hidden within the topological nature of

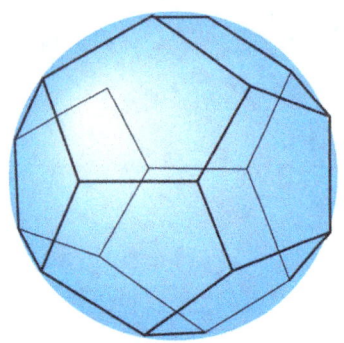

A polyhedron that can be inscribed in a sphere.

the bowl. The key to understanding the deeper meaning of Morse's equation is a discovery made by Leonhard Euler. He noticed that if a polyhedron is inscribed in a sphere, then

$$V - E + F = 2,$$

where V is the number of vertices of the polyhedron, E is the number of edges, and F is the number of faces. This is true for any inscribed polyhedron and thus must belong to the topological nature of the sphere.

Something similar holds for the hemisphere, our open undeformed bowl (which topologically is a sphere with one hole). Inscribe any polyhedron inside this bowl, keeping in mind that it is open at the top. For simplicity, let us take an inverted three-sided pyramid that is missing its base (corresponding to the open top of the bowl). Here we have $V = 4$, $E = 6$, $F = 3$, and therefore

$$V - E + F = 1.$$

The same number was obtained in counting the equilibrium points and their characteristic numbers in Morse's equation. Moreover, we have $V - E + F = 1$ for any such inscribed polyhedron. The equality between the sums,

$$(-1)^{\lambda_1} + (-1)^{\lambda_2} + \cdots + (-1)^{\lambda_n} = V - E + F,$$

is a profound fact discovered by the mathematician Marston Morse about fifty years ago (and is the basis of what is now called Morse theory). Marston Morse (1892–1977), professor of mathematics at the Institute for Advanced Study in Princeton, is famous

An inverted baseless pyramid in a bowl.

for his work in establishing connections between topology and equilibrium points in the calculus of variations. He was one of the first to apply such ideas to minimal surfaces.

But what does all this have to do with minimal surfaces? By the finiteness theorem we mentioned earlier in this chapter, almost all contours behave "reasonably" in the sense that they bound only a finite number of disk-type minimal surfaces. Let C be any such reasonable curve. Then, for some integer N, there are exactly N disk-type minimal surfaces spanning C. We shall label them S_1, S_2, \ldots, S_N. Only some of these may actually be minimal for area.

There is a theorem requiring a great deal of mathematical development that assigns to each of these minimal surfaces S_i a characteristic number λ_i analogous in meaning to the characteristic numbers in our bowl model. In particular, if S_i is of minimum area, then $\lambda_i = 0$. By combining ideas from the fields of topology and the calculus of variations, it can be proved that Morse's equation holds:

$$(-1)^{\lambda_1} + (-1)^{\lambda_2} + \cdots + (-1)^{\lambda_n} = 1.$$

This equation provides a new proof of the previously mentioned result that the existence of two minima implies the existence of a third, unstable minimal surface (see the illustration at the top of page 162 where (C) depicts an unstable minimal surface). To see all this, let us suppose that we had two surfaces of minimum area spanning a reasonable contour and no other minimal surface. Then, because the characteristic number of each minimum is zero, the left side of Morse's equation would read

$$(-1)^{\lambda_1} + (-1)^{\lambda_2} = (-1)^0 + (-1)^0 = 1 + 1 = 2 \neq 1,$$

which is a violation of Morse's equality. So there must be another minimal surface with an odd characteristic number. This other surface thus could not be a local minimum and would have to be unstable. In general, by Morse's equation, the existence of n minimal surfaces of minimum area that span a contour implies that there must be another $n - 1$ unstable minimal surfaces bounded by the same contour.

THE CATENOID AND THE HELICOID

In the preceding sections we considered minimal surfaces bounded by a single closed curve. Let us now investigate minimal surfaces with two or more boundary curves. Such surfaces that have minimum area correspond to soap films bounded by several wires.

What will happen if we dip two closed wires into a soap solution and then withdraw them? We may obtain two separate soap films, one within each wire, or, if we are lucky, we might see *one connected soap film* bounded by both wires, like those shown below. The connected soap films pictured here are of the same topological type as an annulus, which, in turn, is of the same topological type as a sphere with two holes (see the figure C on page 153).

If you try the experiment yourself, you may find it impossible to obtain these surfaces if the wires are too far apart; even if they are close to each other, you may not get a smooth connected soap film, but instead a system of three minimal surfaces, as is shown above on the next page. Such systems of soap films were not on our original list of mathematical surfaces that could possibly span a boundary contour; thus we are once again forced to extend our definition of a surface. In stable equilibria, systems of soap films have the property that only three surfaces can meet at a liquid edge, and they must intersect at angles of 120°. We shall discuss this phenomenon at greater length in the next section.

A soap film spanning two circular wires.

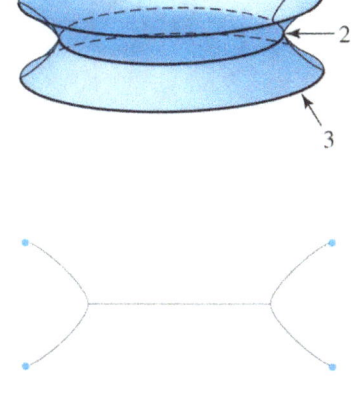

A system of three minimal surfaces with two boundary circles.

The system of three soap films spanning two wires is a relative minimum of area, but it has larger area than the smooth connected annular (ring-shaped) film. It is a simple trick to force the annular surface, and not the system, to be formed in the soap-film experiment. Place a stick (or a finger) through the wires before withdrawing them from the soap solution; then the annular surface will appear when the wires are withdrawn.

There is an amazing relation between this annular minimal surface and a well-known mechanical curve. Let us consider surfaces of revolution. One of the simplest ways to produce a surface is to revolve a curve about a fixed axis. Are there any minimal surfaces that are surfaces of revolution? One candidate is obvious: the plane. It can be obtained by rotating a straight line L about an axis that meets L perpendicularly.

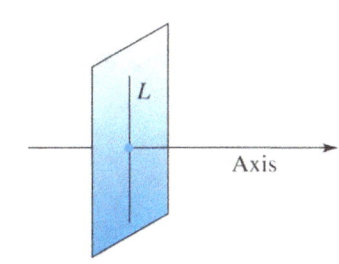

The plane as a surface of revolution.

Are there other surfaces of revolution with vanishing mean curvature besides the obvious planar surfaces, the plane and the disk? There is exactly one other type, and this is how it can be generated. Just hang a chain at both ends from a stick. When it is in rest, you have produced a curve called a catenary.[1] By revolving it about an axis that is

A horizontal stick with a chain, which falls into the shape of a catenary.

[1] The word *catenary* is derived from the Latin expression *catena,* meaning chain or chain line.

A catenoid as a surface of revolution.

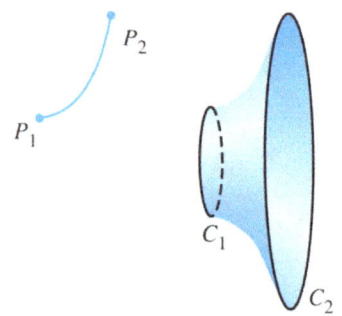

A catenoid bounded by two circles with different radii.

parallel to the stick and below the chain, you will obtain the only kind of nonplanar minimal surface that is a surface of revolution, provided that the lowest point of the catenary is at height $h = \sqrt{b^2 - l^2}$ above the axis of rotation. Here l denotes half of the length of the chain, and b is the distance between the stick and the axis of rotation. This minimal surface, which is the only curved minimal surface of revolution, was discovered by Euler in 1744. For obvious reasons it has been christened the catenoid. As can be seen below, a catenoid can be bounded by circles of different radii.

We have already mentioned that no connected minimal surface will form if the two circles are too far apart. What happens if we start with a catenoid and move the two boundary circles C_1 and C_2 apart? If we perform the corresponding experiment with a soap film, we will find that, at a critical distance between the two coaxial wires C_1 and C_2, the film will break and jump into two disks spanning C_1 and C_2 separately. The distance at which the catenoid must break is known precisely. If, for instance, the circular wires C_1 and C_2 have the same radius R, the critical distance d is about 1.325487 times R.

In fact, the two circles bound two catenoids provided that their distance is less than the critical number d, but only one is stable (the unstable one bends in more), so only one connected soap film can be supported by the two circular wires. The other stable configuration is the pair of disks spanning C_1 and C_2. Thus we see again that the existence of two stable soap films implies that of a third, an unstable one.

With the catenoid we can carry out a very curious experiment. First we build a model of this surface out of some plastic material. We then cut the model along one of its meridians and pull the two cut ends gently apart, thereby twisting the surface, as shown above. In this way we are able to bend the catenoid without any distortion into a piece of another celebrated minimal surface, the *helicoid*. The isometric deformation or bending of the catenoid into the helicoid we have described is even more remarkable because it can be carried out in such a way that all the intermediate surfaces during the bending procedure are minimal surfaces too!

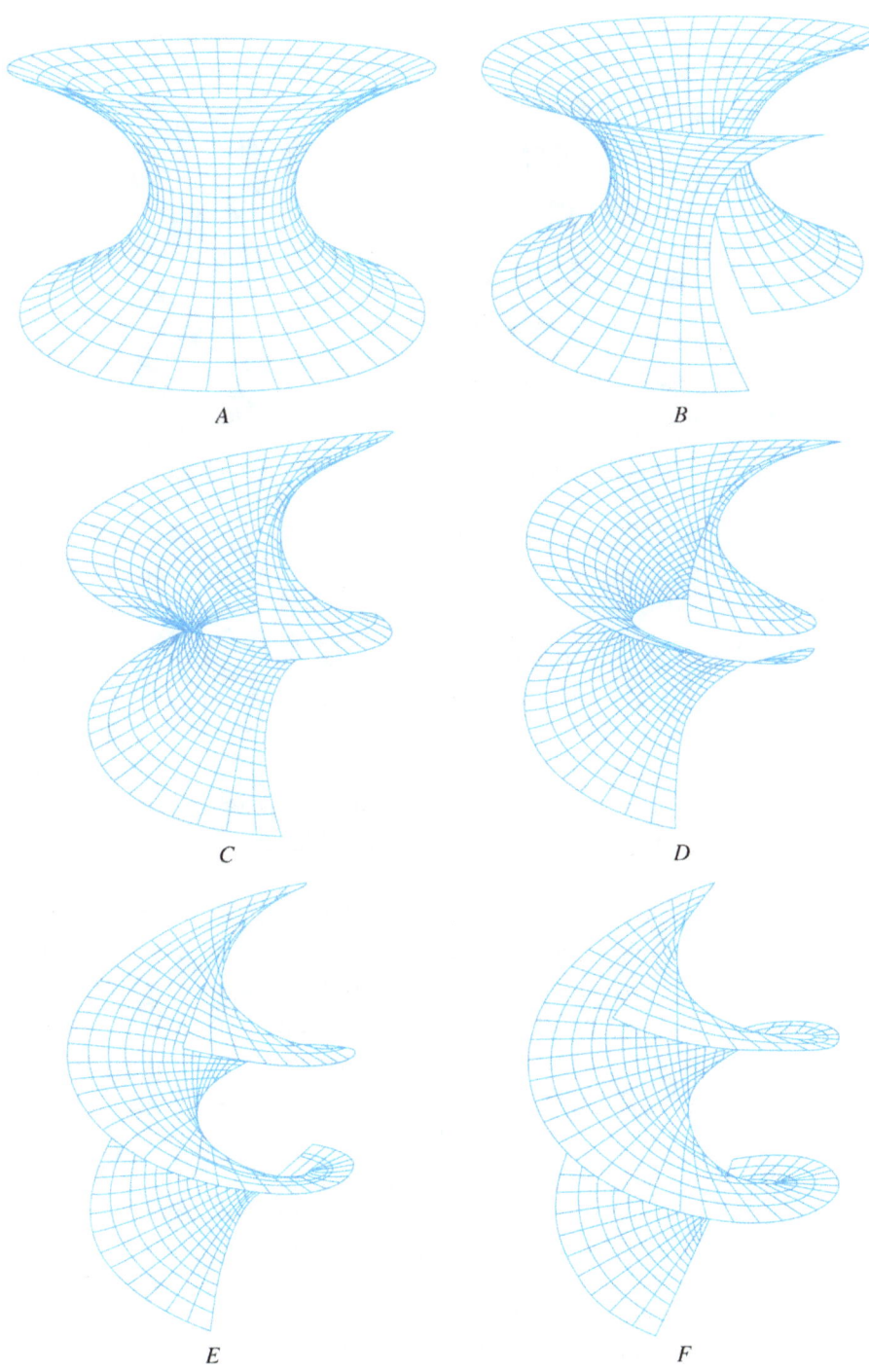

Isometric deformation of a catenoid into a helicoid.

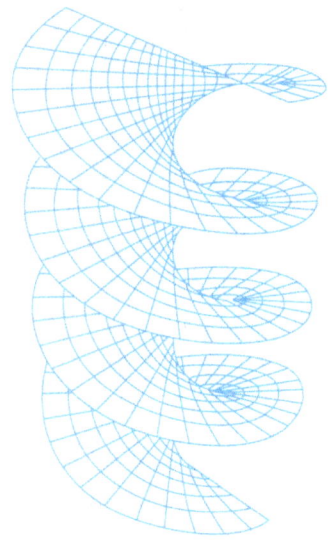

A helicoid, which can be used in the design of winding staircases.

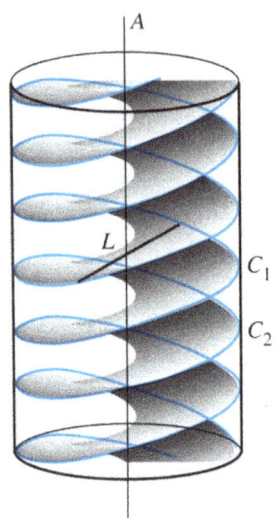

The generation of a helicoid.

Double helix.

A double helix. (Printer's mark, From Aratus of Soli, *Phainomena*, Paris, G. Morel 1559)

After the catenoid, the helicoid, or screw surface, is the second oldest example of a curved minimal surface. It was already well known in 1776, when the French geometer Meusnier discovered that it had zero mean curvature. In fact, architects used parts of the helicoid to build winding staircases, remarkable examples of which can be found in some of the great European residences.

The construction principle of the helicoid is fairly easy to understand. One takes a straight line L that meets some fixed axis A perpendicularly. Then one turns L with constant speed around A and, at the same time, moves the intersection point P of L with A moves with constant speed along A. The superposition of these two motions results in a screw motion of the line L, and the surface swept out by L is a helicoid. The middle figure above shows the part of the helicoid lying inside a cylinder with axis A. The helicoid intersects such a cylinder in two helices, or screw lines, C_1 and C_2, that together form a double helix. Imagine now that each curve is replaced by a thin strip of paper, as in the upper right-hand figure. Then we obtain a geometric figure, the *double helix*, which serves as a model for the DNA molecule, the carrier of the genetic code.

The catenoid is the simplest example of a minimal surface bounded by a configuration consisting of more than one closed curve. Another example is depicted in the upper figure of the next page, where the minimal surface spans two interlocking wires. A system of wires also provides an example of a boundary configuration that can be spanned by infinitely many minimal surfaces of the same topological type. (As we may recall, this is one of the central questions in the study of minimal surfaces.) Let us choose as a boundary a set of three coaxial circular wires of equal

Drawing of a winding staircase (from Hendrik Hondius, *Grondige Onderrichtinge in Optica*, Amsterdam, 1730).

Design of a staircase (from Jacques Androuet du Cerceau, *Leçons de Perspective Positive*, Paris, 1576).

A minimal surface spanning two interlocking curves.

One of infinitely many minimal surfaces that spans a contour consisting of three coaxial circles (below). It is obtained by spanning the configuration of two closed curves depicted above by an annulus-type minimal surface (middle) and then reflecting this surface across the straight arc of the closed curved lying under the circle.

radius and of small but equal distance from each other. Then we can show that there is an infinite family of minimal surfaces of genus zero bounded by these wires. A picture of one such surface is given in the lowest figure at the left. All of the members of the family are obtained by simple rotation of one of them about the axis of the circles. Thus one might argue that this is not a real counterexample to finiteness. This objection is somewhat justified; so we might call two surfaces "really different" if we cannot generate one from the other by a rigid motion in space. The question of how many "really different" minimal surfaces of the same topological type can span a given configuration of 1, 2, 3, . . . curves is completely open.

LIQUID EDGES, AND THE THREE FUNDAMENTAL ANGLES OF 90°, 120°, AND 109°28'16"

Consider the soap film pictured in the upper figure of the next page at the right. This stable minimal surface differs from those we have already encountered. Its boundary does not lie entirely on a given wire but lies partially on a prescribed surface (called a *supporting surface*). Hence the boundary of such a soap film is not, as in the Plateau problem, completely prescribed, but can choose an optimal position on a supporting surface.

To generate this example, we have taken the surface S, made of glass or plexiglass, to which a curve C, a wire, is attached at two points P_1 and P_2. The framework $\langle C, S \rangle$, consisting of the surface S and the attached wire C, is then dipped into a soap solution. When we withdraw it, a soap film attached to the wire C will form; yet some part of its boundary will be a liquid edge on the supporting surface of S. This liquid edge, which we call Σ, is generated by the soap film in such a way that area is minimized. The curve Σ is said to be the *free boundary* of the soap film (or of the stable minimal surface) on the supporting surface S. The upper figure on the next page shows soap films the entire boundary of each of which is a free boundary on the surface S.

We cannot say very much about the shape of the free boundary Σ. In fact, nearly every curve on S can be part of the free boundary of a

A minimal surface which intersects a sphere S perpendicularly (computer simulated solution).

A minimal surface with a partially free boundary.

soap film. Yet one law can be verified in all experiments: smooth parts of a soap film will intersect a smooth supporting surface perpendicularly, as indicated in the middle figure at the right.

H. A. Schwarz gave a rigorous mathematical proof of this fact for minimal surfaces that Plateau had experimentally observed for soap films. Thus we have

■ **Rule 1.** *If a minimal surface has a free boundary Σ on a support surface S, then it meets S along the curve Σ at a right angle.*

This is the 90° rule, which we will meet again and again.

Right angles at the free boundary (a cross-sectional view).

Minimal surfaces with totally free boundaries.

Four soap film experiments demonstrating solutions of partially free boundary value problems.

The oldest problem on minimal surfaces with a free boundary was posed by the French mathematician J. D. Gergonne in 1816:

> Divide a cube by a surface M (shown in gray in the illustration at the right) in two parts in such a way that M is attached at two inverse diagonals C_1 and C_2 that lie on opposite faces of the cube, and M is of minimal area.

The first attempts to answer this question were in error, and the solution was only found by Schwarz in 1872. This minimal surface, though found by Schwarz, is usually called *Gergonne's surface*. It is attached to two other faces S_1 and S_2 of the cube at a right angle.

Boundary configuration $\langle C_1, C_2, S \rangle$ with infinitely many "really different" solutions, formed by helicoids (winding staircases), can be obtained from a cylinder surface S and two perpendicular straight lines C_1 and C_2 that pass through the cylinder axis at different heights.

Gergonne's surface.

The boundary configuration $\langle S, C_1, C_2 \rangle$ of the partially free boundary value problem, the solution of which is shown at the left (computer simulated solution).

 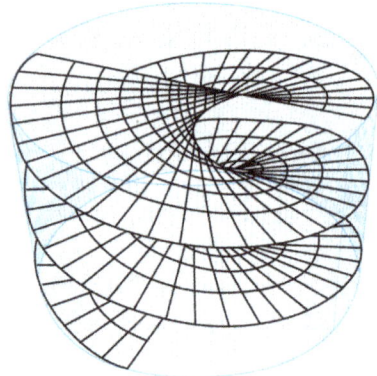

Three of infinitely meany "really different" minimal surfaces that span a certain boundary configuration.

The existence theory for minimal surfaces with free boundaries was begun by Richard Courant in 1940, and in 1951 Hans Lewy carried out the first detailed study of the free boundary Σ. However, Schwarz and the Finnish mathematicians Neovius, Stenius, and Tallquist had already proved the existence of minimal surfaces that span a configuration $\langle C, S \rangle$ for the case where S is a plane and C is a polygon with end points on this plane.

Courant had noticed that we can obtain soap films (that is, minima of area) that span a given frame $\langle C, S \rangle$ and that are really different and are even of different topological type (see the figures below and at the left).

A new situation arises if the supporting surface S itself has some boundary and if the soap film is attached at least in part to the boundary. Then rule 1 is no longer correct, because the soap film is now not completely free to move; the boundary of S forms an obstacle. But the soap film will still meet the "interior part" of the supporting surface S at a right angle. It is interesting to note that, in general, the free boundary of a soap film will attach tangentially to the boundary of S. This has been verified by experiment (see photograph on the next page above, left) and by rigorous mathematical proof.

A free boundary attaches tangentially to the boundary of the supporting surface.

Left: Two solutions of partially free boundary problems.
Right: Cutting holes into the support surface changes the problems into obstacle problems, whose solutions might attach to the boundaries of the holes, which serve as obstacles.

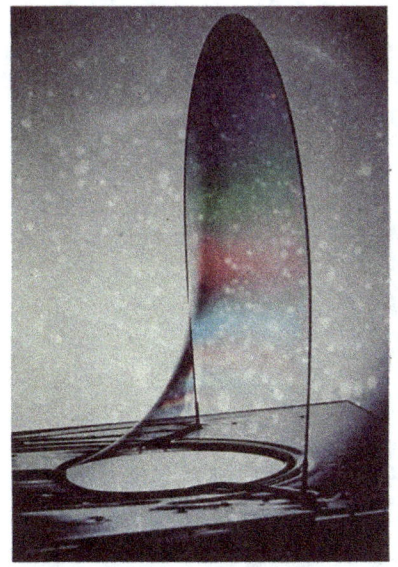

Soap film experiment showing experimental solution of an obstacle problem.

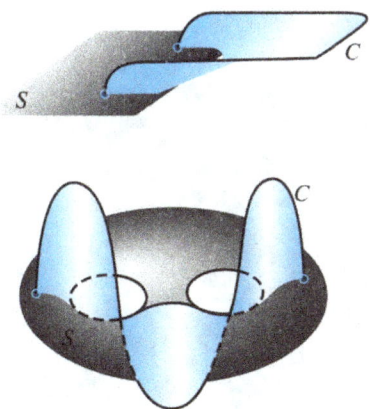

Obstacles on the support surfaces.

188 THE PARSIMONIOUS UNIVERSE

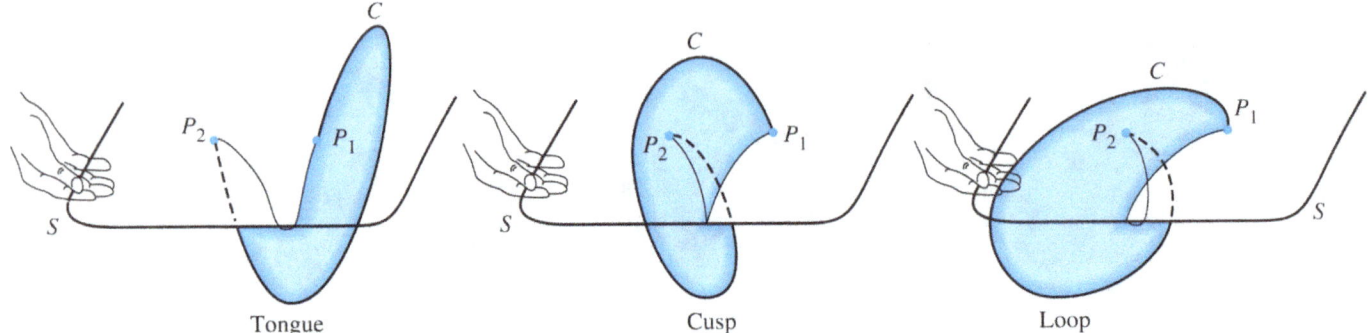

The three configurations—tongue, cusp, loop—that the free boundary of a soap film can form within a configuration consisting of a plate and a wire led around the plate.

However, the free boundary Σ may in some very special cases have a singularity (i.e., a point where no tangent to Σ exists), in the form of a cusp. This can be generated in the following way. Take a glass plate S; fix a wire C at some point P_1 on the upper side of S, lead C around past the edge of S, and then attach it at some point P_2 on the lower side of S (see the figure above). As long as C looks like a circle cut at one point and the two loose ends are moved only slightly apart, the free boundary of the soap film formed has the shape of a "tongue." If, however, the wire C is contorted by deforming it "toward P_2 above the plate S and toward P_1 below S," the tongue will change into a loop. By carrying out this experiment very carefully, you will obtain the cusp as an intermediate position of the wire. The photographs below show the three stages.

Soap-film experiments showing tongue, cusp, and loop.

Another sequence of soap film experiments demonstrating the occurrence of tongue–cusp–loop in partially free boundary value problems for minimal surfaces.

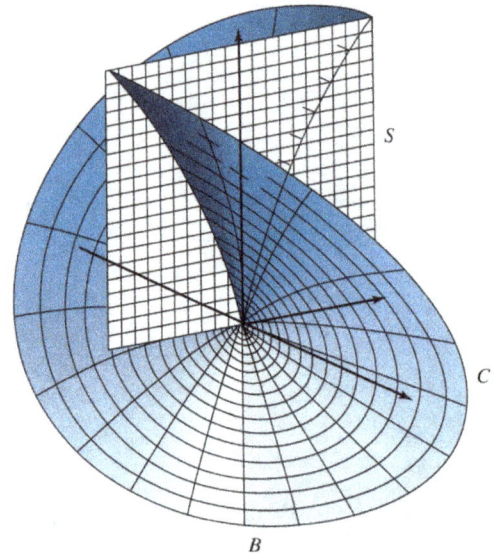

A part of the adjoint of Henneberg's surface which intersects a disk perpendicularly in an asteroid, a four-pointed star. The boundary configuration holding this minimal surface consists of a closed curve and a disk (see the figure left below).

The Henneberg surface.

Furthermore, the figure on the right above depicts a frame, consisting of a half-plane S and a curve C, that bounds a minimal surface, the free boundary of which has a cusp. This surface, which can be modeled by a soap film, was discovered by Henneberg in 1876. Its free boundary is a curve called the *Neil parabola* or *semicubical parabola* (see below). Its two branches can be described by $y = x^{3/2}$ and $y = (-x)^{3/2}$, whereas an ordinary parabola is given by $y = x^2$.

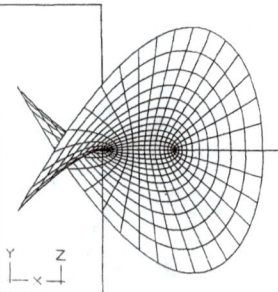

The boundary configuration of the minimal surface depicted left above.

Neil's parabola.

Another view of the part of Henneberg's surface depicted right above.

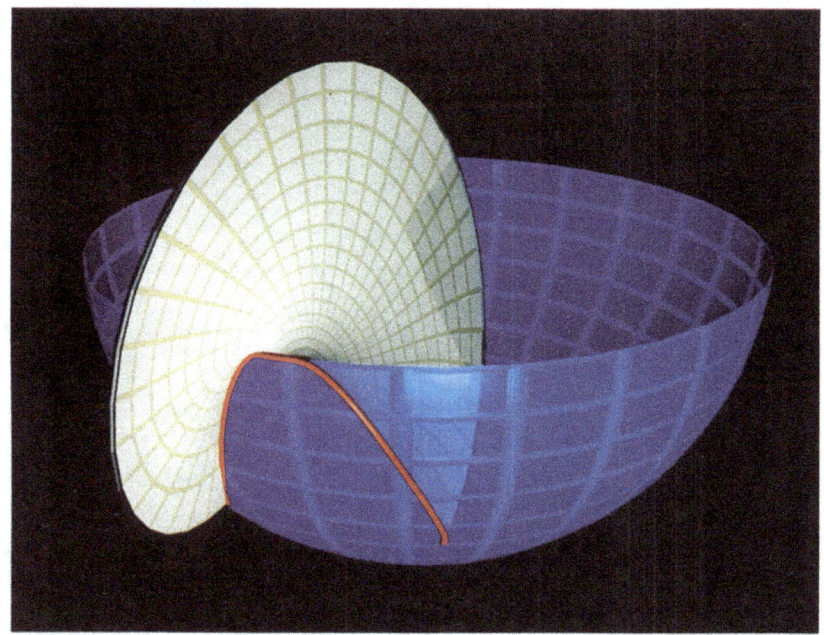

A soap film (bright colour) bounded by an arc (green) and a halfsphere (blue). The film intersects the halfsphere perpendicularly along its trace (red line), except where the trace touches the equator, the boundary of the halfsphere; we see the tongue phenomenon.

Free boundaries on supporting surfaces S are not the only situations in which we can have liquid edges of soap films. Several soap films within a frame can meet at one or more edges to form a stable arrangement of films, that is, a system of surfaces that minimizes area within a given boundary configuration. For example, three arcs C_1, C_2, C_3 meeting in two points (see the figure at the right) can be spanned by a system of soap films with a common liquid edge Σ at which each pair of three films forms an angle of 120°.

Another example is provided by a tetrahedral frame (see the photograph on the next page). It bounds a system of six soap films with four liquid edges. Again, any two films meeting along the same edge must meet at an angle of 120°. Moreover, the four liquid edges have exactly one vertex in common, and each pair of edges meets at an angle φ of 109°28′16″; to be precise, φ is the angle having a cosine of $-1/3$.

Plateau discovered by experiments that these two examples are typical, and nothing else can happen. Thus we have the second rule for a stable configuration of soap films:

A

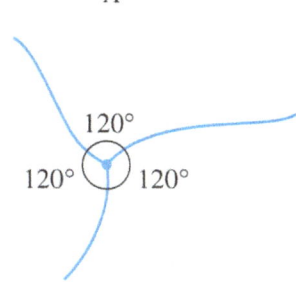

B

A. Three surfaces meeting at 120°.
B. Cross section of their common liquid boundary.

A system of soap films in tetrahedron.

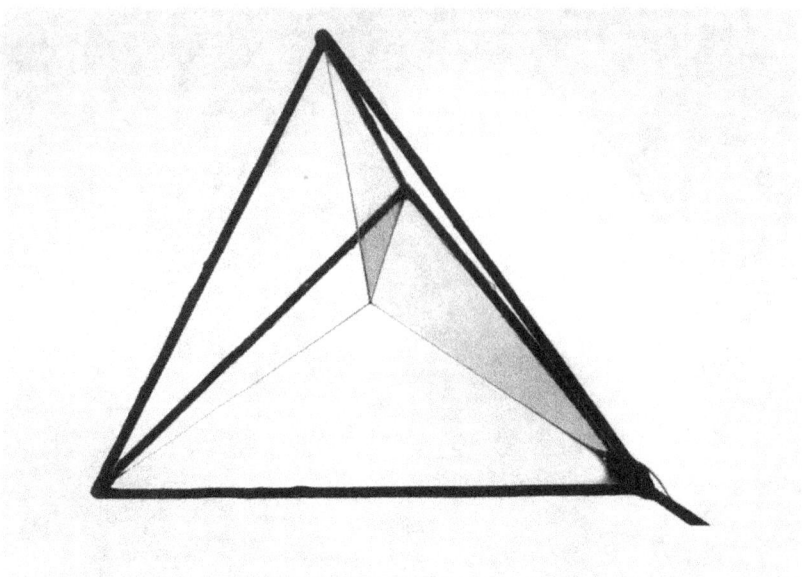

■ **Rule 2.** *Three smooth minimal surfaces of an area-minimizing system of surfaces intersect in a smooth line at an angle of 120°. Only four such lines, each formed by the intersection of three surfaces, can intersect at a single point. At such a point, the angle between any two adjacent lines is 109° 28' 16".*

A rigorous mathematical proof of the part of this rule concerning the 120° has been known for a long time and is rather easy to obtain, provided that the lines of intersection are already known to be smooth. However, a satisfactory proof of the second part was given only recently by Jean Taylor, professor of mathematics at Rutgers University.

The main concept used to prove the second part of rule 2 is the following. Suppose that the minimal surfaces and their lines of intersection are known to be smooth curves. Then we pick some point P from the system of surfaces and magnify smaller and smaller neighborhoods of P to the same size. Thus, in the limit, all surfaces look flat and all lines seem to be straight. The area-minimizing property of the original system of surfaces implies that the new configuration

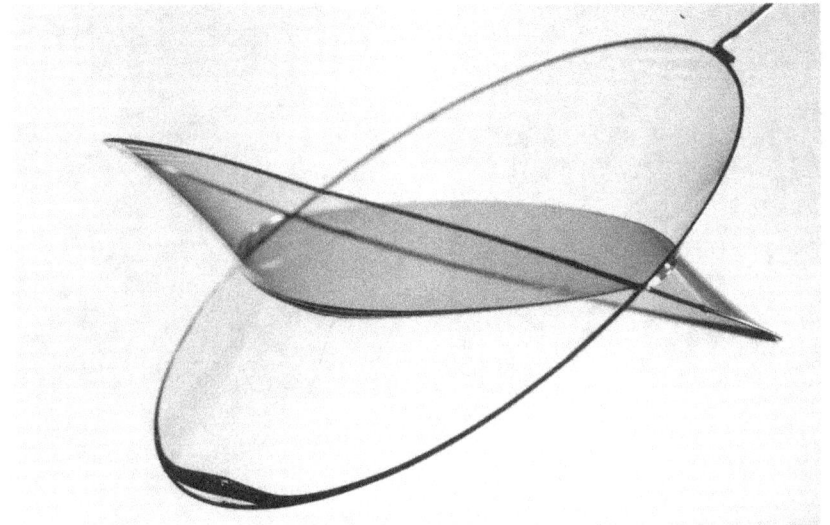

A system of soap films in a boundary configuration consisting of two circular wires.

of the flat surfaces obtained by magnification is also area-minimizing. Now we draw a sphere centered at *P* and let it intersect the flat configuration of surfaces. The intersecting lines are great circles. The area-minimizing property of the flat system means that no more than three of these great circles can meet at a vertex, where they must form an angle of 120°; otherwise the area of the configuration could be reduced.

So the next question is, what are all the possible networks of great circles on the sphere that meet three at a time at equal 120° angles? To answer this, we must solve Steiner's problem on the sphere instead of in the plane. Using spherical trigonometry, we can show that only ten such geodesic networks are possible, and a more detailed analysis shows that only three of the ten are area-minimizing (in the sense that they support area-minimizing cones with a vertex at the center of the sphere). These are the first three parts (parts *A*, *B*, and *C*) in the catalogue of ten depicted on the next two pages. The first one corresponds to a flat surface, and the second two to the systems depicted in the lower figures on page 191 and in the figure at the top of the preceding page. This proves rule 2.

A B C

D E F

G

H

I

There are ten possible networks of great circles on a sphere meeting three at a time at angles of 120°. They are shown here in diagrams A through J. The soap films that span diagrams A, B, and C agree with the surface systems obtained by the "magnification procedure," whereas the soap-film systems in diagrams D through J differ from the corresponding "magnified objects," called "cones," which are generated by all the straight arcs drawn from the center of the sphere to the points on the spherical network. These observations yield a "physical proof" of the fact that the cones for diagrams A, B, and C are area-minimizing, whereas the cones for D through J are not. There is also a rigorous mathematical proof of this fact.

The ten networks A through J can briefly be described as follows: diagram A is a great circle; B is three halves of great circles; C is a spherical tetrahedron; D is a spherical hexahedron; E is a spherical prism with a pentagon as its base; F is a spherical prism over a triangle; G is a spherical dodecahedron; H consists of two quadrangles and eight congruent pentagons; I consists of four congruent quadrangles and four congruent pentagons; and J consists of three quadrangles and six pentagons.

J

The systems of soap films that span polyhedral frames form very beautiful configurations of surfaces. A frame can bound several different configurations of soap films, even several "really different" ones. A very striking example of this phenomenon results from using an octahedral frame. The different soap-film configurations that span this frame can be generated by withdrawing the wire framework from this soap solution in different ways; some experimentation is needed before they can all be physically realized.

A soap-film configuration can often be changed into a different one by gently blowing at the surfaces; but sometimes shaking the frame may work as well.

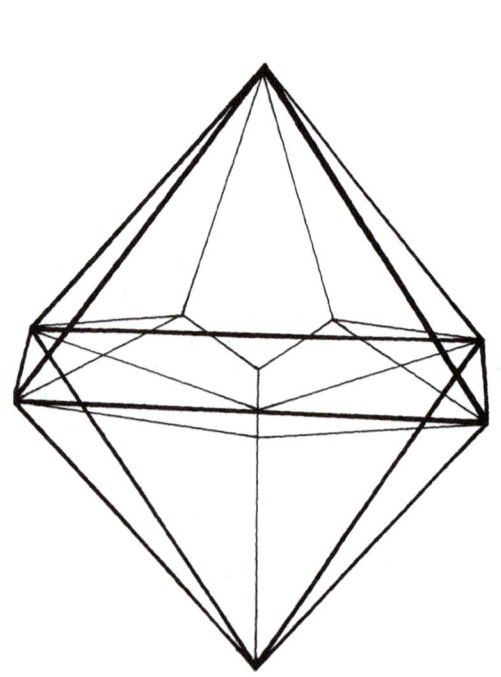

A soap film in an octahedron: frame (thick lines) and liquid edges (thin lines). See also the figure on page 8.

Soap film in an octahedron.

An interesting combination of the 90° and 120° principles can be used to solve Steiner's general problem in the plane experimentally by means of soap films. In this problem we are given *n* points, and we then must look for a network of lines that connects all the given points and has the smallest possible total length.

Suppose we make a frame consisting of two parallel glass or clear plastic plates; these are connected by *n* parallel pins of the same size that meet both plates perpendicularly. If this framework is immersed in a soap solution and withdrawn, a system of planar soap films is formed. These films are attached to the pins, and they have two kinds of liquid edges, both of which are straight lines. One type adheres to the glass (or plastic), which it meets at a 90° angles because these liquid edges are free boundaries on a supporting surface. The other type is where three films meet, forming three angles of 120° (see the figures on the right). Suppose we now mark the position of *n* given points on one of the plates and then fix the pins at these marks. Then the subsystem of liquid edges on either plate will provide an experimental solution of the given Steiner problem, for these edges are subject to the same rules as the Steiner solutions. Moreover, the total area of the soap-film system obviously equals the distance of the two plates times the total length of the liquid edges that lie on one of the plates. Since the soap-film system minimizes area, the subsystem of edges on one plate must minimize length among all connections between the *n* given points.

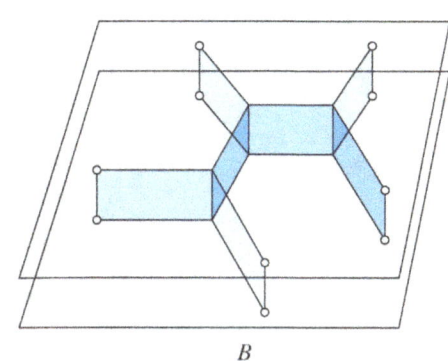

Soap-film solution of Steiner's problem for (A) four points and (B) five points.

PERIODIC MINIMAL SURFACES

H. A. Schwarz was the first to solve Plateau's problem for the simplest contour *C*, which does not lie in a plane. This is a polygon with four edges, the corners of which do not lie in a plane. One calls such a curve a quadrilateral. Schwarz took a very specific quadrilateral, namely a circuit around four of the six edges of a regular tetrahedron. He announced his results in 1865, and he also presented three models he had constructed. The frames of these models were made

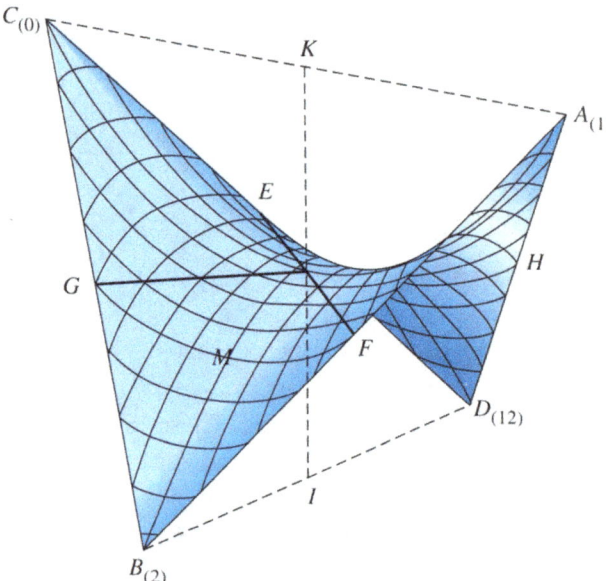

Schwarz's surface in an equilateral frame. Here we have a connected array of four straight lines bounding this minimal surface.

of thin wire, and the spanning minimal surfaces consisted of a gelatinous skin. The first model is shown in the figure above.[2]

Schwarz had discovered that soap films with a free boundary on a surface S must meet S perpendicularly. This led him to consider the following problem. Suppose that $\langle C_1, \ldots, C_k, S_1, \ldots, S_l \rangle$ is a connected array of k straight lines C_1, \ldots, C_k and l planar surfaces S_1, \ldots, S_l, a *Schwarz chain*. This determines a minimal surface that has a boundary on the chain and that intersects the surfaces S_1, \ldots, S_l along its liquid edges at an angle of 90°. Problems of this kind could be explicitly solved in Schwarz's time. Moreover, Schwarz discovered two important *reflection principles for minimal surfaces*: reflection of a minimal surface across a straight line and reflection across a plane. Both reflection principles allow us to build up "large" minimal surfaces from small pieces. These reflection principles can be formulated as follows.

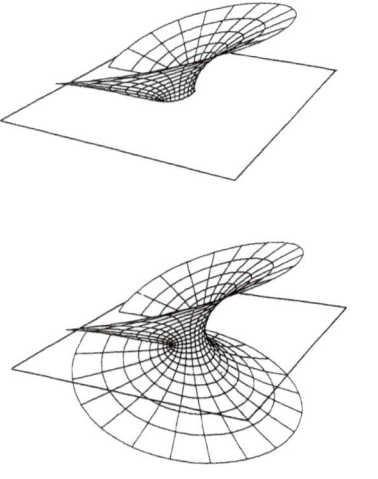

Reflection of a minimal surface across a plane.

[2]The paper describing this work was presented to the Berlin Academy in 1867 and was awarded the Academy's prize. Schwarz's paper is still one of the marvels of mathematical literature. Even a modern reader may enjoy it because of its beautiful style and its flawless presentation. Bernhard Riemann had also solved the problem in 1861–1862 but did not publish his results. It was his student Hattendorff who revised Riemann's manuscript in 1866 and published it posthumously in 1867.

■ **Rule 3.** *If a part of the boundary of a minimal surface M is contained in a straight line, then its reflection M* across the line is also a minimal surface, and the union of M and M* forms a smooth minimal surface as well.*

■ **Rule 4.** *If a minimal surface M meets a plane at a right angle, then its mirror image M* across the plane is a minimal surface that, together with M, forms an extended surface that is a smooth minimal surface.*

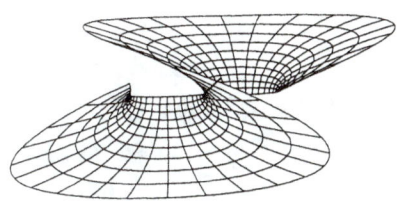

Reflection of a minimal surface across a straight line.

Consider now a minimal surface M whose boundary consists of several pieces that lie either on straight lines or on planes that meet M perpendicularly. Then we can reflect M across each straight piece of the boundary, and we can also form all the mirror images of M across those planes that M strikes at a right angle. In this way we get a new surface; it consists of the original surface M, the *basic unit*, and of its reflections and mirror images, which we generated by employing the reflections described in rules 3 and 4. The new surface is a smooth minimal surface, and its boundary (like that of all reflections and mirror images of the basic unit M) has the same properties as the boundary of M. That is, its boundary consists of finitely many pieces that lie either on straight lines or on planes that meet the surface at right angles. We therefore can proceed with this process of extension: each reflection and mirror image of the basic unit can be considered to be a new basic unit to which the two reflection rules may be applied.

A part of Schwarz's periodic surface.

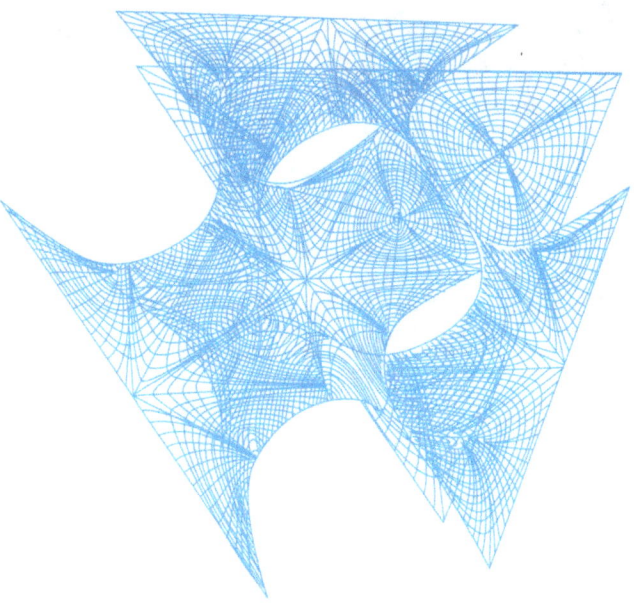

The extension process will terminate if at some stage no boundary is left at which a reflection can be carried out. For instance, if the basic unit is a half-plane, only one reflection is possible. One can, however, prove that *for a finite basic unit the extension process will never come to an end.*

We therefore can expect to arrive at an "infinitely extended" minimal surface. We must, however, anticipate that this surface will in general have *self-intersections*. Only very special Schwarzian chains bound minimal surfaces that, considered to be basic units, lead to an *infinitely extended*

200 THE PARSIMONIOUS UNIVERSE

(a)

(b)

(c)

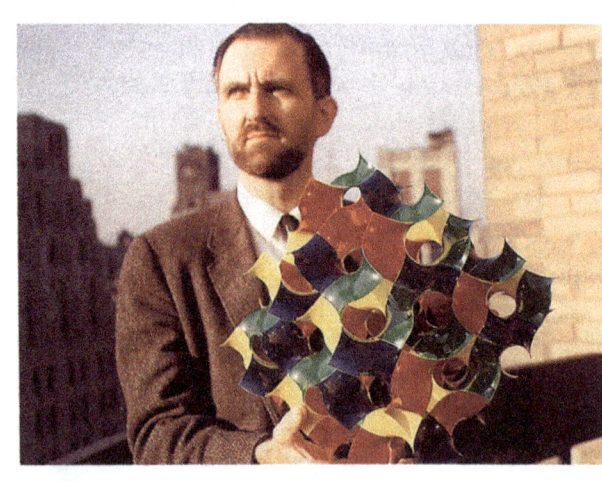

(d)

(a–c) Parts of Schwarz's second periodic minimal surface (models produced by Alan Schoen). (d) Alan Schoen with a model of his gyroid, another periodic minimal surface.

minimal surface without self-intersections; such an object is called a *periodic minimal surface*.

As early as 1867, Schwarz pointed out that the (uniquely determined) minimal surface spanning the quadrilateral in the figure on page 198 is the basic unit of a periodic minimal surface.

Schwarz and other authors found additional periodic minimal surfaces, some of which are represented in the photographs above and

(e)

(f)

(g)

(e) A model of Schoen's gyroid.
(f) A polyhedral approximation of the gyroid. (g) A periodic minimal surface of genus 29 per fundamental region (model by Alan Schoen).

on the facing page. The models depicted in our photographs were designed and prepared by the American physicist Alan H. Schoen in 1968–69.

Periodic minimal surfaces and related structures might be interesting for biologists, who have noticed that separating walls between organic and inorganic material in the skeletons of echinoderms (starfish, sea urchins, and related creatures) resemble certain kinds of periodic minimal surfaces.

Interior view of the Olympic Stadium.

Roof of the Olympic Stadium (Munich, 1972).

MINIMAL SURFACES AS ROOFS AND TENTS

In the past thirty years, the elegant constructions of the architect Frei Otto and his collaborators have justifiably become well known. Their buildings are not constructions in the classical sense of architecture but are more like an arrangement of tents. For instance, the German Pavilion at the 1967 World Exposition in Montreal, designed by Frei

Exterior view of the Olympic Athletic Arena (Munich, 1972).

Otto and Rolf Gutbrod, resembles an unearthly landscape formed by a single membrane of varying height. Its contours are supported at the high points by masts, and at the low points the membrane is anchored to the ground. The prestressed membrane, consisting of a transparent material, is supported by a steel wire net, which is connected to the mastheads and anchor points by a system of edge ropes, ridges, and eye loops. The "eye loops" (the importance of which we shall explain later) are filled with clear plastic.

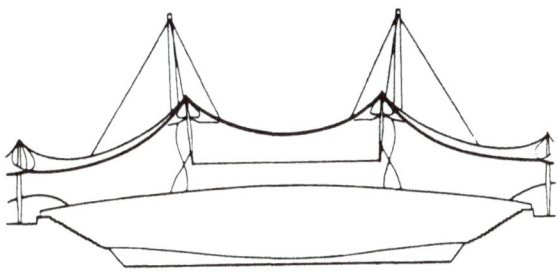

Roof for the Olympic Athletic Arena (Munich, 1972) showing five main units of minimal surfaces, eight eye units, two suspension points, two central secondary points, eight support points, and twenty-nine anchor points. Maximum span, 443 feet; maximum height, 180 feet.

Another celebrated design is the roof of the Olympic Stadium in Munich. This is also a tentlike construction formed by saddle-type nets with two suspension points, two support masts, various anchor points, and a single frontal-edge cable 1,443 feet in length. The net is covered by a translucent acrylic sheeting. The steel masts are 190 feet tall, and the roof has a maximal span of 213 feet. The roofs of the Olympic Athletic Arena and of the Olympic Swimming Arena are two other constructions of the same type.

The similarity between tents and the roofs of Frei Otto is not an accident; they have several principles of construction in common. First, both are conceived as lightweight structures, designed to be economical by using the least possible amount of construction material. Second,

Roof for the Olympic Swimming Arena (Munich, 1972) showing a minimal surface with one suspension point and thirteen stabilization and anchor points. Maximum span, 279 feet; maximum height, 164 feet.

Exterior view of the Olympic Swimming Arena (Munich, 1972).

they are designed to be easily erected, dismantled, and transported. Third, the principal force acting on these roofs is *tensile stress*; only a few parts, such as the masts or supporting arches, are exposed to *compression*. (One finds stress in tight ropes or in a spanned tent, while in pillars or walls one finds compression.)

Tents and lightweight roofs thus differ in an essential way from most conventional buildings, in which the amount of material used is generally out of proportion to the actual load requirements of the building. Ordinary buildings cannot be moved to other locations without great difficulty. Moreover, most of the materials used in them, such as stone or bricks, can be subjected to compression forces, but not to stress. The bending and buckling forces in

these buildings are usually dealt with by using a large amount of extra material, leading to inevitable waste and unnecessary cost.

As we have already seen, nature provides a handy device for detecting optimal shapes of surfaces within a given contour: just permit a soap film to span a desired boundary configuration. If it does not easily break, it will be in stable equilibrium; it is a minimal surface that minimizes area. Frei Otto and his collaborators have used soap films as a principal tool for their architectural designs. They have carried out many experiments to find elegant forms that can be transformed into real buildings.

Exterior view of the Olympic Swimming Arena (Munich, 1972).

Most of these experiments were in principle already known to Plateau and other researchers, but Frei Otto's ingenious variations of the geometric design of the supporting boundaries of the soap films led to marvelous new forms that had not been conceived of before.

One type of material Frei Otto used in these experiments is hair-thin thread tied to the ends of needles or thin sticks, which in turn are fastened to holes in a plexiglass plate (see the photograph below). If we dip such a configuration into a soap solution and then withdraw it, the soap film will pull the threads taut in order to achieve a position

A soap film bounded by a system of threads.

Threads on a disklike minimal surface; (A) loose thread; (B) thread tightened.

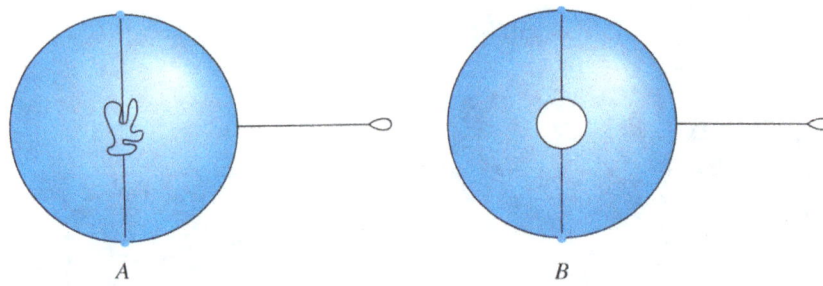

of minimal area. By fixing the threads to needles of different height, we obtain a spectacular tentlike shape.

For these soap-film models to be transformed into architectural forms, they must be carefully photographed and measured. Solid models are then produced and tested in wind tunnels. Special measuring devices inform the experimenters about the stress that might occur under the impact of wind and snow. Then, in the actual construction, thin, high-strength steel cables replace the hair-thin thread used in the soap-film models; the membranes are usually made out of synthetic material.

Very interesting aesthetic effects can be achieved by the following procedure, which you can easily carry out yourself. Take a wire in the form of a planar curve, attach a handle to it, dip it into a soap solution, and then remove it to form a (planar) soap film. Then place a thin loop of thread onto the film, where it will lie crumbled and without tension. With a blunt instrument and some delicacy, you will be able to break the soap film within the loop without destroying the film outside the loop. The soap film wants to reduce its area; so it pulls the thread tight, into the shape of a circle (see the figures above). A circle formed in this way is shown on the next page above. Here the thread is not attached by strings to the exterior curve as shown above, but one of its points P is tied to a needle, which is fastened to the ground. If we lower the needle, as in the photograph on the next page, the thread is pulled out of its planar position and becomes a spatial curve, which, incidentally, will be a very smooth curve of constant curvature, except for the point P where the needle is attached and those parts of the loop that cling together. The soap

A planar minimal surface bounded by a circular thread.

A thread under vertical stress (pulled downward).

film bounded by its exterior contour and the "eye loop" is now a nonplanar surface, and its curviness will provide a stable form for possible later architectural use. For example, the photographs on page 209 show how Frei Otto's institute in Stuttgart was conceived. In the model, an exterior contour was formed by threads tied to needles that are inserted into a plate. Moreover, a gallowslike upper needle supported a loop to which the soap film was attached. In the

A thread pulled upward forms an "eye loop."

building, steel cables have replaced the threads, and the interior of the loop has been filled with a clear plastic material, which, together with the windows on the ground floor, floods the interior with daylight.

One can now easily understand the principles of design and construction that Frei Otto used on the roof of the Olympic Stadium in Munich and on the German Pavilion in Montreal. Here we find a whole system of supporting masts (instead of a single gallows) and, accordingly, many eye loops. The results are tentlike roofs of great elegance. In addition to the supporting eye loops, other supporting elements, such as ridges, humps, or ropes, are also used (see the figures on page 210).

It is well known and easily verified that we can push a soap film with a blunt object such as a finger, but not with a needle, which will pierce the film. On the other hand, using the edge of a sharp knife, or a very thin thread, we can lift a soap film. This raises the questions of what objects can be used to lift a soap film, and exactly what a *set of coincidence* (the set of points where the object touches the soap film) looks like.

Problems of this kind, called *obstacle problems*, have been an important and fruitful field of mathematical research and have led to the

The Institute for Lightweight Structures, Stuttgart: (upper left) model; (upper right) network; (right) the finished building.

Generalization of Steiner's problem: a minimal system of roads in a landscape with a lake.

development of many new and interesting techniques in the calculus of variations. It is obviously useful to study such problems. If, for instance, we plan to connect several cities by a road system of the shortest total length, but the roads must avoid some lakes, mountain areas, and national parks, we are faced with a generalization of Steiner's problem that must take obstacles into account (see the figure at the left). The solution of optimality questions under such subsidiary conditions turns out to be one of the most important problems in applied mathematics. We met another type of obstacle problem when we considered minimal surfaces with a free boundary.

Many problems related to the Frei Otto or "thread" experiments have been solved experimentally but not mathematically. For example, what is the highest point that a soap film can reach if it is lifted from a planar contour by an eye loop (see the top figure on the left of the next page)? We have seen that two curves C_1 and C_2 cannot be spanned by a connected soap film if they are too far apart. In fact, C_1 and C_2 cannot bound such a minimal surface if the two curves can be separated (as in the figure on the left) by a circular cone that is generated by revolving a pair of straight lines, meeting at a right angle, about an axis of symmetry. Given the length of the eye loop and the position of the exterior contour C, we can thus easily derive an upper bound for the height above the ground of the minimal surface in the top figure on the left of the next page; yet it would be interesting to find the exact value of the maximal height.

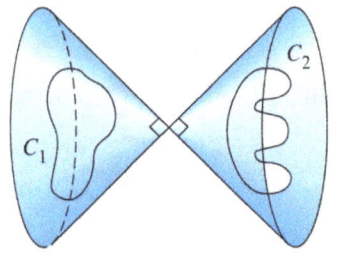

Two cones give a nonexistence result.

Obstacle problems formed by structural elements of light-weight buildings.

Humps: view from above

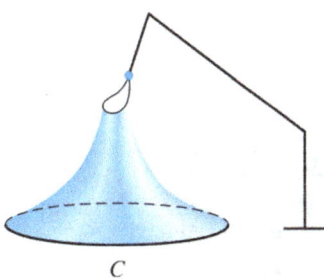

How high can an eye loop lift soap film?

Supporting ropes

Minimal surfaces supported by ropes, an obstacle problem. (Structural elements of light weight buildings.)

Top view

Side view

Top view

Side view

Minimal surfaces with several eye loops attached to one point.

We could also attach several loops to the same point (see the figures above), or fix a thread at the end points of a given arc. A soap film will then span the thread in such a way that it minimizes area, and the thread will form a curve of constant curvature where it is not attached to C. The photograph on the right shows a soap film that is bounded in part by a wire in the form of a helix and in part by a thread tied to some points of the helix.

In this chapter we have become familiar with a variety of geometric forms and topological shapes of liquid films, all the result of the interplay between the surface tension of these films and the structure of their boundary configurations. It seems quite surprising that the phenomenon of capillarity can create such a variety of forms, even in one of its simplest manifestations, that of soap films. In the next chapter we begin by investigating what happens if we place air pressure to one side of a soap film, and then we shall turn our attention to some other equilibrium configurations, particularly those of liquids.

Helix with thread.

Section of a glass-painting of two boys blowing soap bubbles (Cologne, about 1530). Wallraf Collection, Schnütgen-Museum, Cologne. The painter has illustrated the antique proverb "homo bulla" (man is like a bubble), which Erasmus had included into his *Adagia*, a collection of proverbs.

6 Optimal Design

THE ISOPERIMETRIC PROBLEM

In the fifth century A.D., the Greek philosopher Proclus wrote, "The circle is the first, the most simple, and the most perfect figure." The perfect symmetry of the circle justifies this statement, as does the beautiful property found by Queen Dido—that among all planar figures of equal perimeter, the circle (i.e., the circular disk) has the maximum area. This property of the circle is often called the *isoperimetric property of the circle*.[1]

[1] This comes from Greek (*isos* = equal, *peri* = around, round about, *metron* = measure) and alludes to the fact that we look at figures whose circumference has the same measure, that is, the same length.

Nature does nothing in vain, and more is in vain when less will serve; for Nature is pleased with simplicity, and affects not the pomp of superfluous causes.

(Isaac Newton, *Principia*)

We can express this remarkable fact by a kind of dual formulation, characterizing a circle by a minimum property instead of a maximum one:

> Among all planar figures of equal area, the disk has minimal perimeter.

In order to derive this minimum property of the circle from Queen Dido's maximum property, we derive a very interesting inequality. To this end, consider an arbitrary closed curve C of perimeter L. Suppose that it encloses a domain of area A. Let r be the radius of a circle of perimeter L; that is, $L = 2\pi r$. Such a circle encloses a disk with an area of πr^2. By the isoperimetric property of the circle, the area A enclosed by a curve C cannot be larger than πr^2, and it equals πr^2 if and only if C is a circle. On the other hand, we have

$$\pi r^2 = \frac{1}{4\pi}(2\pi r)^2 = \frac{1}{4\pi} L^2,$$

and we thus obtain the celebrated *isoperimetric inequality*:

$$A \leq \frac{1}{4\pi} L^2,$$

between the perimeter L of an arbitrary closed plane curve and the area A enclosed by it, and where the equality sign holds only for the circle. From this inequality it easily follows that, among all planar figures of equal area, the disk has minimal perimeter. To see this, consider an arbitrary planar figure with perimeter L and area A. Let D be the disk with the same area A and the perimeter l; then $l^2 = 4\pi A$. It follows that $L \geq l$, because $L < l$ would imply the inequality $L^2 < 4\pi A$, violating the isoperimetric inequality. Actually, one can see from this that the "minimal" and "maximal" characterization of the disk are two equivalent formulations of the isoperimetric property of the circle. The "minimum" characterization also explains why the oil drops on top of your broth are circular, not triangular or hexagonal. The molecular forces generate a figure of least circumference—that is, of smallest potential energy for a given amount of oil—and this is a disk. If two floating oil drops hit each other, they quickly merge into a single, larger oil drop.

Not surprisingly, there are several other optimum properties of the circle. For instance, *among all planar domains of a given area, the disk can support the largest sand pile.* This may prove to be valuable information for your children or grandchildren if they like to build sand castles.

A very similar problem arises in the theory of plasticity: *What is the cross section of a perfectly elastic column which can withstand the largest torsional moment (that is, the largest amount of twisting)?* Again, the cross section must be circular.

Another beautiful feature of the disk was observed by the British physicist Lord Rayleigh. It concerns the musical sound generated by a membrane, for example, by beating a drum. As you may know, one can lower the principal frequency of the membrane by "increasing its size," say, by increasing its area. It would then be interesting to learn which membrane, among all planar membranes with a given area, has the lowest principal frequency; that is, *which drum, among all drums with a given cross-sectional area, has the lowest tone?* By experimentation, Rayleigh (1877) was led to conjecture that a circular membrane has this minimum property, but it took almost half a century until this conjecture was proved by Faber and Krahn (1923–1924).

A similar statement was proved for metal plates (like a cymbal) whose boundary is "clamped," that is, fixed by some appliance:

> Among all clamped plates of given area, the disk-shaped one has the minimal principal frequency.

In fact, the field of acoustics provides many other interesting applications for the calculus of variations. We can even say that the theory of pure tones, the very field of Pythagorean physics, is nothing but a subfield of variational calculus.

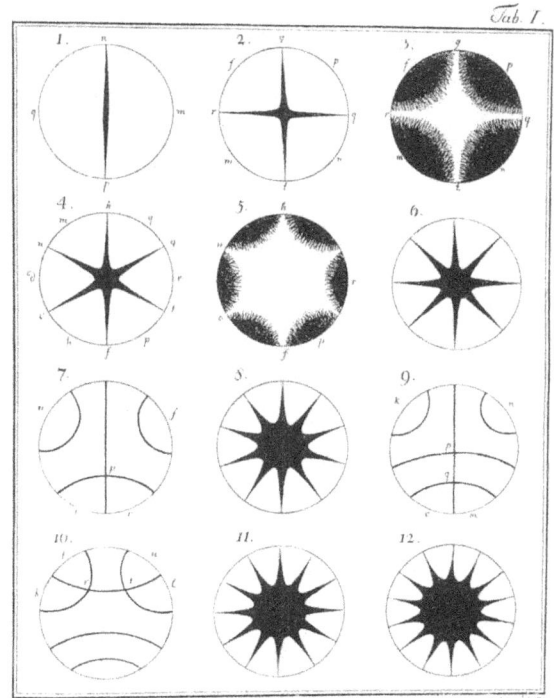

Chladni figures of vibrating plates and bells. (Engravings taken from E.F.F. Chladni, *Entdeckungen über die Theorie des Klanges*, Leipzig, 1787.) Some of the plates are horizontally supported, otheres are clamped; numbers 3 and 5 show motin of water in a vibrating bell.

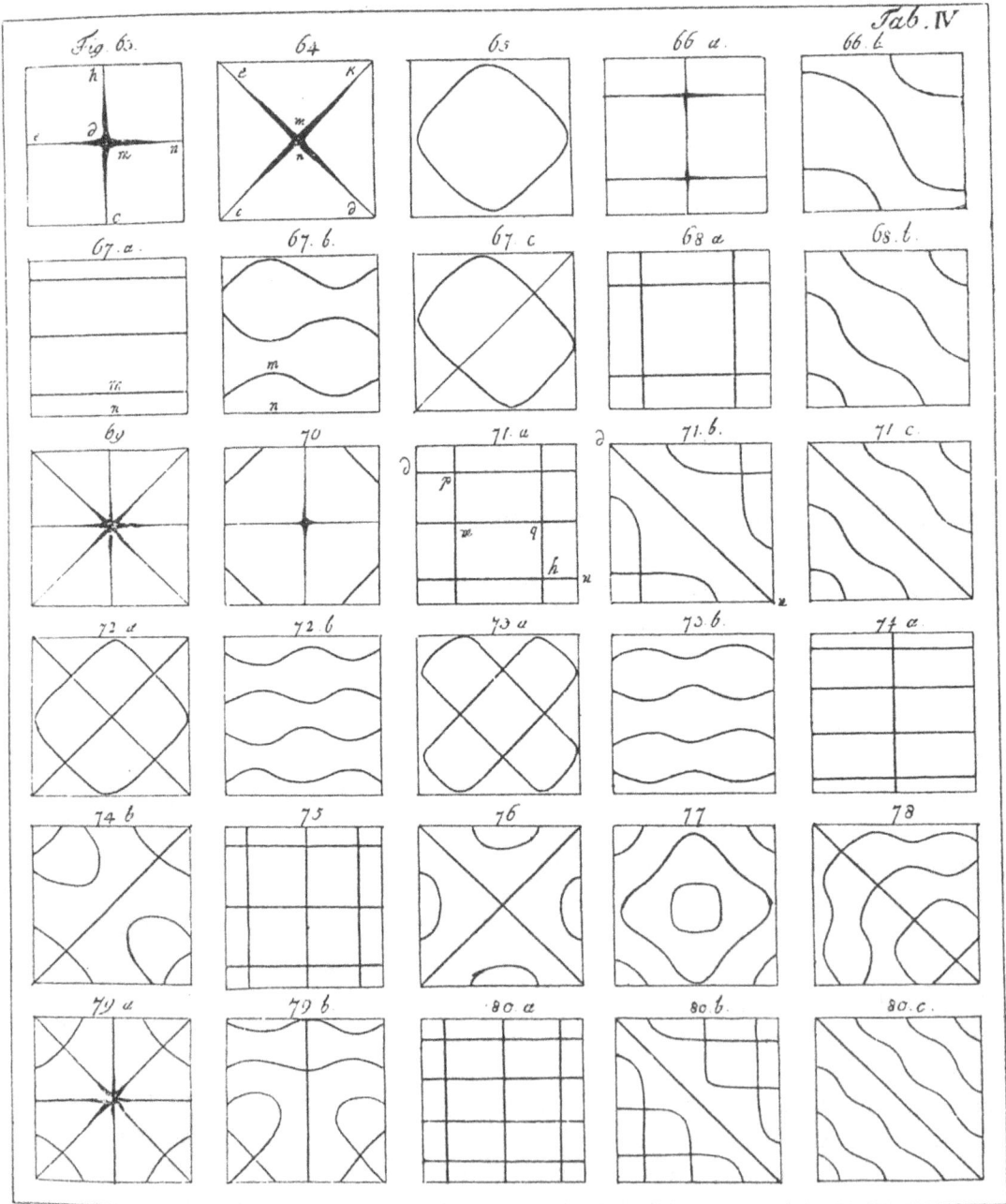

hladni is one of the founding fathers of acoustics. His famous treatise *Die Akustik* appeared in 1802. Chladni spread and on plates made of metal or glass which were fixed in clamps. He then applied a violin bow to the edge of such lates, and the sand grains gathered at those places at rest then created the nodal lines. It is still a largely unsolved roblem to make theoretical predictions about size and shape of the nodal lines of an arbitrary plate. Therefore the esign and construction of musical instruments, say violins, is an art and not a science.

A PSEUDOPROOF OF THE ISOPERIMETRIC PROPERTY OF THE CIRCLE

We have, however, not yet seen a proof of the isoperimetric property of the circle; so let us look at the reasoning that Jacob Steiner gave in 1836.

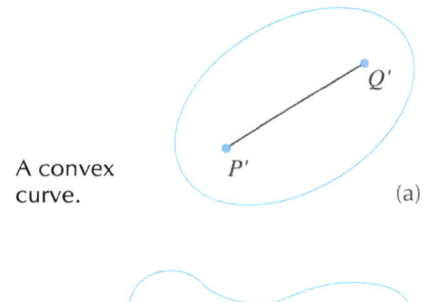

A convex curve. (a)

Assume that a solution of the isoperimetric problem exists. Then we have *a curve C that, among all closed lines of a given length, encloses the maximal area.* We would like to show that this curve is a circle. First, we note that it is a convex curve; that is, every straight connection of two points P' and Q' inside curve C lies is entirely within the region enclosed by the curve [compare parts (a) and (b) of the illustration on the right]. Otherwise curve C would lie on one side of a straight line, L, that passes through points P and Q on the curve, so that those points of L that are between P and Q do not belong to curve C [see part (c) of the illustration]. Then, as indicated in part (d), we could construct a new curve C^*, which would enclose more area than curve C but would have the same perimeter. This, however, contradicts our assumption that C encloses maximal area. Hence our solution curve must be convex.

A nonconvex curve. (b)

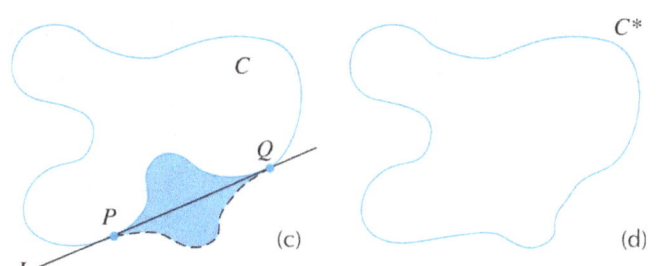

If C is not convex, then by reflection along L we generate a new curve C^* as shown in part (d). This new curve, although of the same length as curve C, encloses more area than does C.

Next, we choose two points R and S on the convex curve C such that these points divide C into two arcs of equal length, C' and C''. The straight line through R and S cuts the interior of C into two pieces, B' and B'', as shown below at the left. The optimum property of curve C implies that B' and B'' have the same area. This can be seen as follows. Suppose, for example, that B' had a larger area than B''. Then, if we reflected region B' in the line segment RS, we would obtain a new region, the mirror image of B'. The union of B' with its mirror image would then form a figure that had a larger area than that of the interior of C, whereas its perimeter would equal the length of C. This, however, would contradict our assumption about the optimality of curve C; so the areas of B' and B'' must be equal.

Mirror image of B'

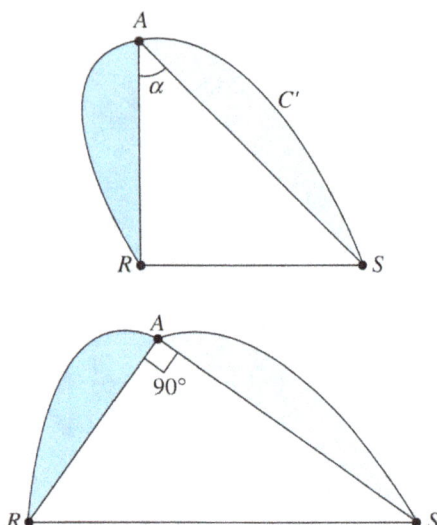

Using a mechanical argument to complete the proof of the isoperimetric property of the circle.

To prove that C is a circle, it suffices to show that both C' and C'' are semicircles. Once more, we will argue by contradiction. Suppose that one of the two arcs—say, C'—were not a semicircle. Then a classical theorem attributed to Thales would imply that there is a point A on arc C' such that the angle α of triangle RAS at corner A is *not* 90°. Imagine now that sides AR and AS could be moved by a hinge installed at A, so that we could reduce or increase the angle at the joint, and, furthermore, that the two lunules (crescent-moon forms) shown in color in the illustration at the left were rigidly attached to the arms of the hinge, so that they moved with them when they opened or closed. We could then change the opening of the hinge from the original angle, α, to an angle of 90°, thereby increasing the area of triangle RAS. (Clearly, the two points R and S at the tips of the two lunules will have a different position from before, if the angle of the "scissors" with hinge A is changed.) Together with the attached lunules, the new figure would have a larger area than the old one with angle α at A. Yet as we have seen, the area of the old figure was half that of the interior of curve C. Therefore, if we reflected the new figure in the straight line through R and S and unified it with its mirror image, we would obtain a domain with the same circumference as C but with a larger area. This cannot be; so arc C' has to be a semicircle. It follows in the same way that C'' is a semicircle, and therefore C is a circle.

We thus have proved that the circle is the only solution to the isoperimetric problem. For Steiner, the problem was solved at this stage. Reportedly, his colleague Dirichlet tried without success to convince him that the proof was not complete. As we have seen, the existence of a solution of a reasonable geometric problem is by no means obvious; in fact, sometimes a solution may not even exist. For the isoperimetric problem, however, the existence of a solution can be verified. Thus, *the only curve that maximizes area among all closed curves of the same perimeter is the circle.* We will not present the existence proof because it is beyond the scope of this book. Consequently, our proof of the isoperimetric property of the circle, remains incomplete and is therefore just a "pseudoproof."

BALLS AND BUBBLES

For mathematicians, a ball is a solid whose surface is a sphere. It is a perfectly round body that looks the same from all sides. For a homogeneous ball, the geometric center coincides with the center of gravity. Therefore, on a horizontal plane, the equilibrium position of such a ball is always undecided. It can rest in all positions; yet the slightest kick, in any direction whatsoever, will cause it to roll. Furthermore, we are never sure what path it will take, since a tiny unevenness of the ground could make it turn. Thus the ball became one of the attributes of the capricious goddess Fortuna. "She is the one so reviled even by those who should praise her but who, instead, give her ill repute. Yet she is blessed and does not hear; happy with the other primal creatures she rules her sphere and rejoices in her bliss" (Dante, *The Divine Comedy, Inferno*, Canto VII, "The Avaracious and the Prodigal," verses 91–96).

Another symbol of the fleeting fortune is the soap bubble. This airy creation is, as long as it lasts, a consequence of Bernoulli's law of virtual work. This law, as we have seen, requires that in stable equilibrium the potential energy has to be a minimum. In a soap bubble, some amount of air is enclosed by a surface of minimum area, the soap film. This is the physical evidence for the mathematical theorem that *out of all solids of a given volume, the ball has the smallest surface area*, a result that is the obvious spatial analogue of a minimum property of the circle: among all planar figures of a given area, the circle has the shortest perimeter.

For the same reason, globules of oil suspended in a liquid of the same density (or brought into outer space, which is nearly free of gravitational forces) will form perfect balls. This can be seen in experiments, and the corresponding mathematical theorem has been rigorously proved.

In 1919, the Swedish mathematician Torsten Carleman showed that the ball is the only equilibrium figure for a self-gravitating liquid at

Albrecht Dürer's "Small Fortune."

The square seat of Virtue and the rotund seat of Fortuna. Woodcut from Charles de Bouelles, *Opera*, Paris, H. Stephanus, 1510.

Le bulles de savon. Etching by Jean Jacques de Boissieu (1799).

rest. In huge masses, the powerful forces of self-attraction predicted by Newton's law dominate other forces, such as surface tension, which can then be disregarded. Let us suppose that the planets arose from liquid masses that hardened as they cooled. The ball is then the only shape that a planet at rest could have achieved. The situation is different if the celestial body is rotating, as do the Earth, the other planets, and most of their satellites. We shall consider this situation later on.

Recall that a ball of radius r has a volume of $V = (4/3)\pi r^3$ and a surface area of $A = 4\pi r^2$, from which we obtain the relation $36\pi V^2 = A^3$. We can then employ an argument similar to the proof of the isoperimetric inequality to derive the inequality

$$V^2 \leq \frac{1}{36\pi} A^3$$

between the volume V of an arbitrary solid and its surface area A, where equality holds only for balls. This estimate is the spatial version

Glass blowing.

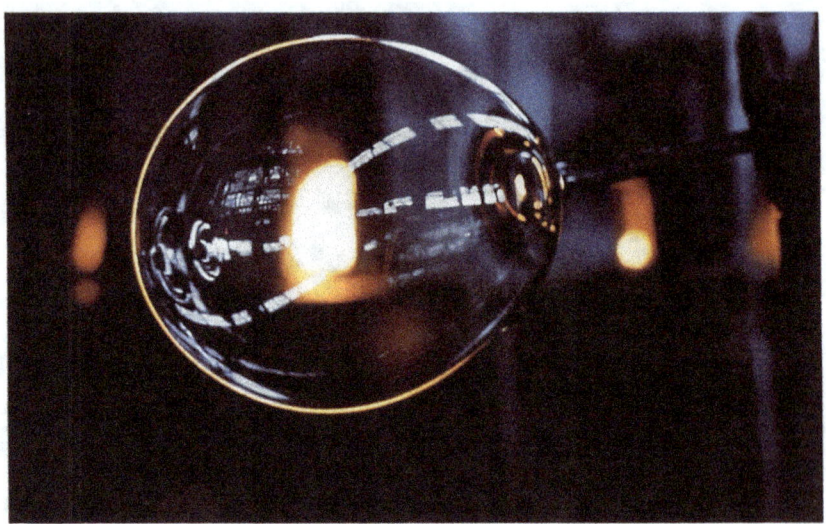

of the isoperimetric inequality. Therefore, in analogy with the circle we have the following maximum property of the ball:

Of all solids with prescribed surface area, the ball has the largest volume.

Three soap bubbles with the same boundary contour.

We can alter the problem of finding surfaces of least area enclosing a prescribed volume by adding suitable boundary conditions. For example, we may consider the problem of finding the solid of largest volume with a prescribed surface area when the part of the boundary touching some given plane is not counted. The solution is a half-ball, as we can prove by a reflection argument.

The following problem is closely related. Consider a solid wedged between parallel planes. The *reduced surface area* of such a body consists of those parts of the surface that do not lie on either plane. *For a given value of reduced surface area, what is the solid of maximum volume wedged between two planes? Equivalently, what wedged solid of a fixed volume has the least possible reduced surface area?*

We can solve this problem experimentally by blowing a soap bubble between two wet glass plates. Let us begin with a bubble in the form of a hemisphere that sits on one of the wet plates. As we blow more air into the hemisphere, we enlarge it until it also touches the second plate, whereupon it changes into a circular cylinder that sits perpendicularly on both plates and meets them in circles, as shown in the illustration at the right. Then, by the same reasoning used for the "experimental" solution to Steiner's problem, we can conclude that the circle is the shortest curve that encloses a given area. As we have seen, this fact is equivalent to the isoperimetric property of the circle.

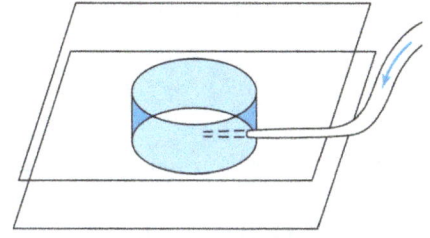

Soap-film demonstration of the isoperimetric property of the circle.

We can use a plate experiment to establish an interesting connection between the Steiner problem and the isoperimetric problem. Suppose that we have two glass plates connected by three parallel pins of equal length that meet both planes perpendicularly. Then, using a straw, we blow a cylindrical soap bubble between the two wet plates in such a way that the cylindrical soap film contains all three pins as shown in the illustration below.

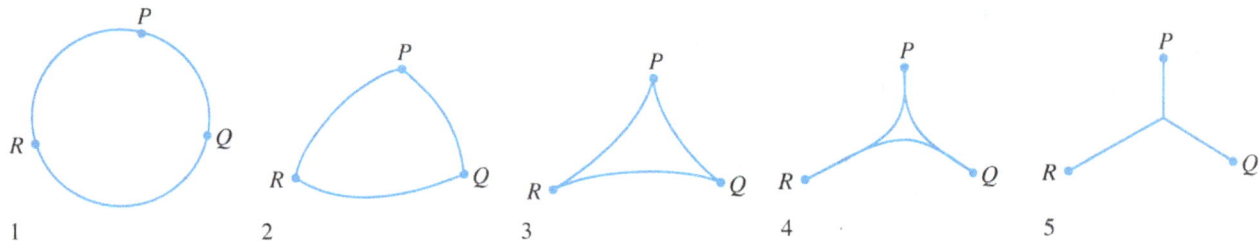

A cylindrical soap bubble between two plates held apart by three pins of equal length demonstrates the relationship between Steiner's problem and the isoperimetric property of the circle.

Then we slowly draw air out of the bubble. The cylindrical shape will continuously change through a sequence of configurations, and it will eventually become a system of three soap films that meet both glass plates perpendicularly. The liquid edges of the soap-film configurations at various stages during this procedure are illustrated above; this, as we see, connects the isoperimetric with the Steiner problem. Note that in each configuration the liquid edges are either straight or circular arcs, because only planar and cylindrical soap films will arise.

A near dodecahedron within a bubble cluster, created by Tom Noddy.

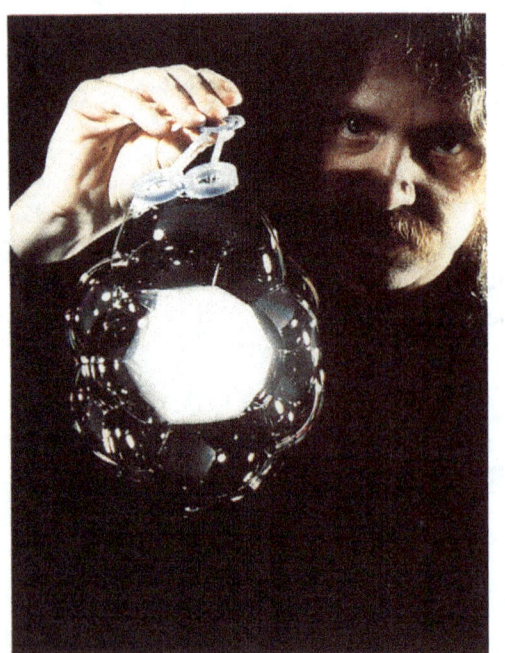

A nearly cubical bubble within a bubble cluster, created by Tom Noddy.

Now let us put circular cylinders of the same size on a plate and try to arrange them as closely packed together as possible. Then we will obtain a hexagonal pattern, because this provides the tightest packing of circular objects of the same size in the plane (see the upper illustration in the margin, in which the central circle touches exactly six neighbors). This interesting "close-packing" property is another reason why the hexagon, the 120° angle, and the Y-shape so often occur in nature.

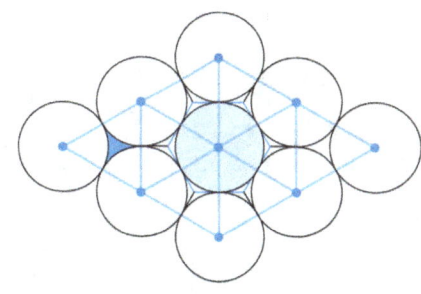

Similarly, densely packed balls of equal size between two plates arrange themselves in a hexagonal array. Imagine each ball to be a living cell that tries to expand as much as possible and each by the same amount. Then it seems obvious that a pattern of hexagonal cells will form. In fact, this is very often what we see with cell growth. Exterior pressure may lead to the same result if it is uniformly applied.

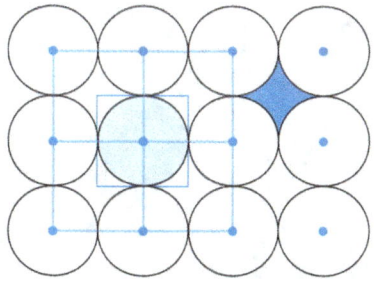

However, the hexagonal pattern can also be obtained in a completely different way. Let us blow soap bubbles of the same size between two parallel plates. The bubbles will soon unite. According to rules 1 and 2 for systems of soap films, which apply to this situation as well, we then obtain a system of soap films with liquid edges that obey the 90° and the 120° laws (see Chapter 5, pp. 183 and 192). Moreover, if the bubbles are the same size, there is no pressure difference between two adjacent cells; so all interior films must have mean curvature zero—in fact, they are planar surfaces. Hence we get an array of hexagonal cells where only the outermost films are cylindrical, because for them there is a pressure difference between the outside and the inside of the cell (see the photograph at the right).

Tightest packing of circles. The hexagonal packing of disks (say, coins) minimizes the space left between them relative to the space left in a square array (as in the lower drawing).

Packing of soap bubbles between two parallel plates.

We see that the hexagonal pattern may form for different reasons; that is, the minimal-area configuration can be generated by very different forces, such as uniform pressure or surface tension. *Therefore, it is impossible to deduce from the observed patterns alone which forces are acting.*

Nevertheless, as the naturalist D'Arcy Thompson pointed out in *On Growth and Form*, scientists have made exactly such careless deductions when trying to explain the geometric structure of the beehive. As we all know, one of the most beautiful hexagonal arrays is the honeycomb constructed by bees. The writings of Pappus inform us that the ancient Greeks had already tried to explain the regularity of beehive cells by means of an optimum principle. The French physicist R. A. F. deRéaumur (1683–1757) also thought that the hexagonal structure of the bee's honeycomb should follow from a minimum principle: the bee would build its cells with the greatest economy in order to use as little wax as possible. De Réaumur posed his conjecture to Samuel Koenig, the later adversary of Maupertuis. Koenig found that angles of 120° and 109° followed from the minimum principle proposed by de Réaumur, and these seemed to agree with the actual measurements of the bee cells. Thereupon Fontenelle, the secretary of the

Yüncü–Kilim (19th century, Northwest Anatolia, Turkey).

Honeycomb.

French Academy, declared that bees had no intelligence; yet they were "blindly using the highest mathematics by divine guidance and command." On this D'Arcy Thompson commented that it makes more sense to suppose "that the beautiful regularity of the bee's architecture is due to some automatic play of the physical forces" than to suppose "that the bee intentionally seeks for a method of economizing wax."

Cross section of a honeycomb.

But all this assumes that the bees *have* somehow hit upon the optimal honeycomb. Have they? This question was investigated by the Hungarian mathematician Fejes Tóth in 1964. In his paper "What the Bees Know and What They Don't Know," he considered *honeycombs*, which he defined as a set of congruent convex polyhedra, called *cells*, filling the space between two parallel planes, without overlapping and without interstices, in such a way that (1) each cell has a face (called a *base* or *opening*) on one and only one of the two planes and (2) every pair of cells is congruent in such a way that their bases correspond to each other.

Longitudinal section a of honeycomb.

The cells built by the bees are prismatic vessels, the openings (and cross sections) of which are regular hexagons, whereas their bottoms consist of three equal rhombi (see the illustration at the right.) The bees construct their honeycomb in such a way that the hexagonal openings of the cells are attached to one of the two planes. Is the zigzagged bottom surface constructed by the bees the most economical one? (It is certainly more advantageous than a plane.)

In order to state the problem precisely, we formulate (following Fejes Tóth) the *isoperimetric problem for honeycombs*:

An individual bee cell (lid).

> Given any two numbers V and W, find a honeycomb of width W whose cells have the smallest surface area and yet enclose the volume V.

(The width W is the distance between the two parallel planes that bound the honeycomb.)

We don't know yet what the solution will be, but it definitely cannot be the bee cell, because Tóth found another cell that yields a slightly

Fejes Tóth's cell.

better result. The bottom of this cell consists of two hexagons and two rhombi (see the illustration to the left). Tóth's cell saves less than 0.35% of the area of an opening (and a much smaller percentage of the surface area of a cell). Hence we can state that the bees do a pretty good but not perfect job, although their practical result, taking the margin of error into account, might still be optimal.

Let us return to a consideration of the many configurations of soap films, soap bubbles, or combinations of soap bubbles that span frames of various kinds. We can use wires, glass plates, or surfaces

Two bubbles on a plate.

Four bubbles.

A "worm" of seven bubbles.

of liquids on which the bubble configurations will rest or float. We can also combine supporting surfaces, curves (wires), and movable lines (threads). Some possible configurations are shown on pages 228–231.

A "worm" of seventeen bubbles produced by Tom Noddy, whose record stands at eighteen bubbles.

A conglomeration of bubbles, or foam.

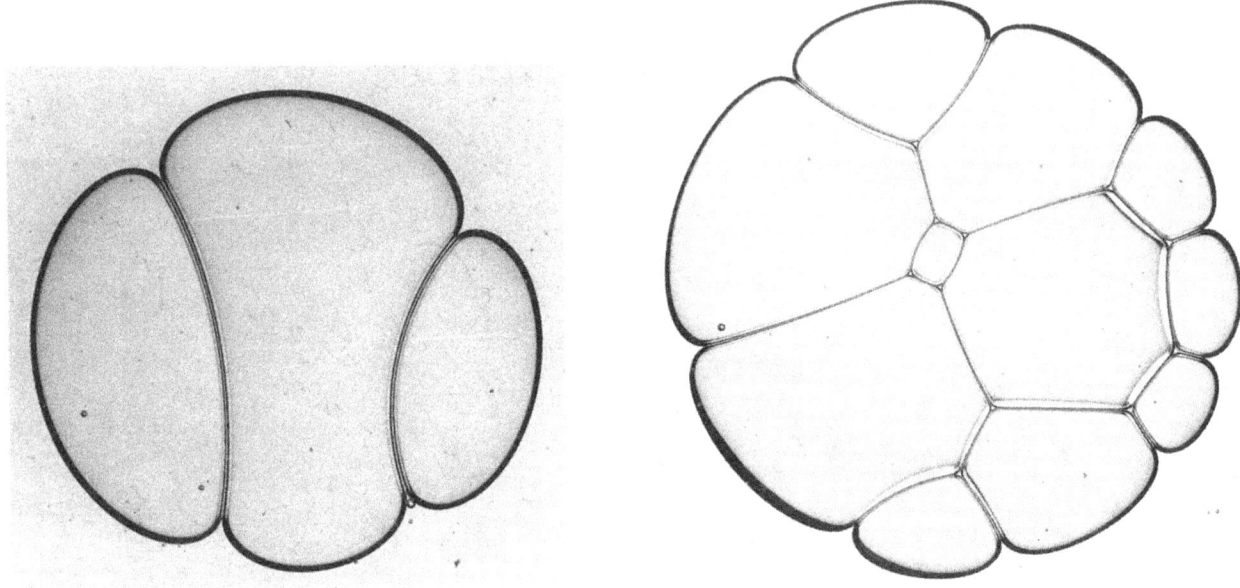

Cross-cuts of conglomerates of soap bubbles between two paralled plates and vertical walls (black lines as cross-cuts).

A cluster of soap bubbles.

Bubbles in threads, a plaster model.

A diagram of threads, holding bubbles.

Consider the mathematical problem of describing soap bubbles or configurations of soap films and soap bubbles. As we already know, the problem is to minimize surface area among all configurations consisting of surfaces that span a given framework (rigid or movable) and enclose some prescribed volume. The solutions to such a mathematical problem consist of one or more surfaces of constant mean curvature. Minimal surfaces are those whose mean curvature is zero, whereas spheres and cylinders have constant mean curvatures different from zero. *Thus the solutions of the mathematical model problems are conglomerates of surfaces with different mean curvatures.* They are models of the soap-film/soap-bubble arrays observed in experi-

ments. Minimal surfaces correspond to soap films exposed to the same pressure on both sides, whereas surfaces whose mean curvature is nonzero correspond to soap films exposed to more pressure on one side than on the other.

Although such mathematical surfaces are a good approximation to physical reality, they should not be identified with it. It is essential that we distinguish the physical phenomenon from its mathematical model, particularly because the mathematical concept may encompass a much larger realm of objects than we can actually realize by experiment. If, for example, a surface of constant mean curvature lacks stability, it cannot be physically embodied in a soap film. We have already seen that there are unstable minimal surfaces (for example, certain catenoids) that cannot be physically embodied in soap films. Similarly, the cylinder's surface becomes unstable if its length equals or exceeds its circumference, which is π times its diameter. This can be easily demonstrated by an experiment that does, however, require some delicacy (see the legend on the next page). Consider a cylindrical soap bubble between two coaxial rings. As soon as the cylindrical film becomes unstable, it will decompose into two separate, spherical soap bubbles of different sizes. The instability of long cylinders can be not only demonstrated by experiment, but also proved rigorously in the mathematical model.

The cylindrical soap bubble's instability, which was discovered by Plateau, is closely related to a similar physical phenomenon with which everyone is familiar: every jet of water breaks into a series of water droplets. We can observe this phenomenon at a fountain or by watching a sprinkler in a garden. The dispersion of water jets into a spray of droplets is very sensitive to perturbations like those caused by electrical fields or sound vibrations. In his classic book, *Soap Bubbles, Their Colours, and the Forces Which Mold Them*, the British physicist C. V. Boys describes many simple experiments that can be done with water jets or soap bubbles. Boys' book grew out of three lectures given in 1889–1990 by this outstanding experimentalist, who occasionally enjoyed shooting enormous smoke rings from the windows of his laboratory at startled pedestrians in an effort to surround them by the rings.

Let us now again illustrate the difference between reality and a mathematical model by considering a seemingly simple example, the spherical soap bubble. Nature has no difficulty showing that spheres are the only smooth soap bubbles (without an artificial boundary) that you can blow into the air. However, until 1984, we did not know whether spheres are the only complete finite surfaces of constant mean curvature. (The unlimited cylinder, also a complete surface of constant mean curvature, is not finite; it extends to infinity in both directions.) Most mathematicians conjectured that the answer

(A) Soap bubbles between two coaxial rings. They usually are not cylinders but have developed a waist or, as in the pictures above, a bulge. (An explanation of this phenomenon is found on pages 231-233.) To obtain cylinders from the bubbles with a waist or a bulge, we must either add or extract air.
(E) A soap bubble joining two rings in a plane. (The two rings are also covered with spherical caps.)

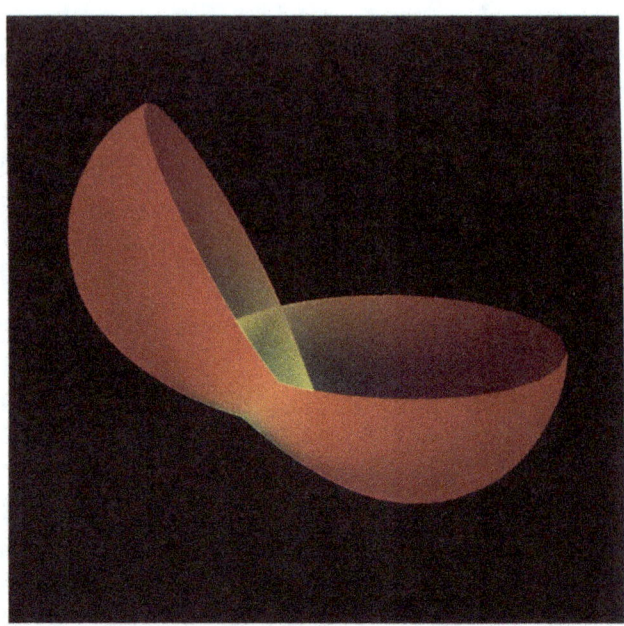

H. C. Wente in 1984 discovered a closed finite surface of constant mean curvature and genus 1. In these computer-generated pictures, Wente's surface is composed, step by step, from three building blocks. The finite closed Wente surface (without boundary) does not reveal how the self-intersections are generated. To understand the construction, one should look at the sequence of four pictures from left to right. The two figures on this page show a building block in two different positions.

Closed curve (*A*) without self-intersection and (*B*) with it.

was, "Yes, spheres are the only ones," and various partial results were proved. We knew, for example, that spheres were the only *convex* surfaces of constant mean curvature among the finite complete surfaces. A. D. Alexandrov even proved that a non–self-intersecting handlebody surface of finite genus (greater than zero) cannot have constant mean curvature. Yet mathematicians wanted to obtain a completely general result. For this purpose, surfaces with self-intersections had to be taken into consideration also. Admitting such surfaces, Heinz Hopf was able to show that finite surfaces of genus 0 and without boundary must be spheres, even if they have self-intersections. Recently, however, the conjecture has been disproved: for example, there exists a complete finite surface of genus 1 and of constant mean curvature, but it is self-intersecting.

Two building blocks put together.

The final Wente torus consisting of three building blocks.

ROTATIONALLY SYMMETRIC SURFACES OF CONSTANT MEAN CURVATURE

(*H*-SURFACES OF REVOLUTION)

Surfaces of constant mean curvature are the models of soap films, soap bubbles, and perhaps other forms appearing in nature. In order to determine natural forms described by *H*-surfaces, it is, in particular, very useful to find out which of these surfaces are also rotationally symmetric, like the disk, the catenoid, and the cylinder. We shall call them *H-surfaces of revolution*. Plateau experimentally found that there are exactly six different kinds of *H*-surfaces of revolution: the plane and the catenoid (mean curvature nonzero), the sphere, the cylinder, the unduloid, and the nodoid (mean curvature zero).

Closed surface (A) without self-intersection, and with it (B).

Cylinder Unduloid Sphere Catenoid Nodoid

Circle Ellipse Straight line Parabola Hyperbola

The generating curves of the *H*-surfaces of revolution. The roulade of a circle generates a cylinder, that of an ellipse generates an unduloid, and so forth.

Several *H*-surfaces of revolution (cylinder, unduloids, sphere).

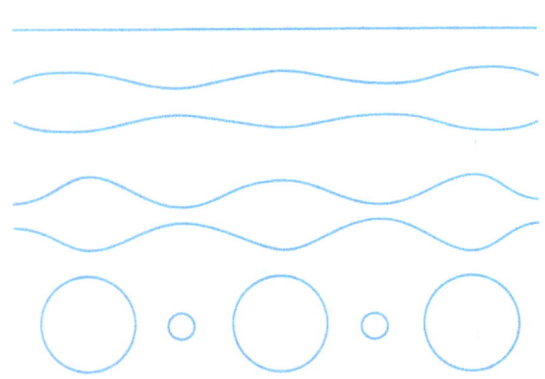

Earlier, in 1841, the French mathematician Delaunay had proved that this list of *H*-surfaces of revolution is indeed complete and that, moreover, all of them can be generated by a simple rule, which can be described as follows. We fix some straight line to serve as the *axis of rotation*. Thereafter, we choose a conic section—that is, an ellipse, a circle, a parabola, or a hyperbola—and we imagine that the conic section rolls on the axis. Each focus of the conic section will then describe a curve called the *roulade* of the focus. When rotated in space about the axis, this curve generates one of the six *H*-surfaces of revolution. (Five of the generating curves are shown at the left.) For example, the focus of a circle describes a straight line parallel to the axis, and the corresponding surface of revolution is a *cylinder*.

If an ellipse rolls on the axis, either of its foci generates an undulatory (wavy) curve that never touches the axis, and this curve, when rotated, generates the *unduloid*. If the foci of the ellipse are not very far apart, each of the undulatory curves is only slightly wavy and, if the foci move together, the unduloid more and more closely approximates a cylinder.

The opposite extreme of an ellipse is a straight line segment, which can be thought of as a degenerate ellipse in which the two foci are as far apart as possible: the two end points of the segment are its foci. The roulade of this line segment is a sequence of semicircles (one end touches down, the other swings around, tracing a semicircle, and so on), which upon rotation generates a sequence of equal *spheres*.

If one focus of the ellipse remains finite but the other moves out to infinity, the ellipse changes into a parabola. The roulade of its focus is a catenary, which is, as we have seen, the meridian of the *catenoid*.

Finally, we will describe how the roulade of one of the foci of a hyperbola is generated. Imagine that the two branches of the hyperbola are rigidly connected, so that they can move only as a unit. Roll one branch of the hyperbola on

the upper side of the axis, say, to the right, from one asymptote to the other,[2] and then roll the other branch on the lower side of the axis to the right, from one asymptote to the other. By repeating the rolling again and again, always moving to the right, and by following the trace of one of the foci, we obtain a curve that extends infinitely to the right. In a similar way, this curve can be extended infinitely to the left. Thus we obtain a smooth curve with infinitely many loops, which generates the *nodoid* as the surface of revolution.

The *plane* is obtained by rotating a straight line around an axis that intersects it at a right angle. Yet such a perpendicular is seen to be the "limit" of a sequence of semicircles whose radius is increasing toward infinity. Thus the perpendicular can be interpreted as the roulade of an infinitely long and infinitely thin ellipse.

One cannot help but be reminded here of the ancient Greeks' construction of new curves by letting well-known curves roll on other curves, as discussed in Chapter 2.

HOW TO TAKE A BATH IN A SPACESHIP

As we have seen, the principle of virtual work implies that soap-film configurations consist of surfaces of constant mean curvature. Furthermore, from this principle it follows that free surfaces of liquids not under the influence of gravitation are such surfaces.

How can we switch off gravitation? We can either shoot a container with liquid into outer space or we can place it in an elevator, which is then dropped in free fall. The latter experiment has been done and has led to an impressive confirmation of various mathematical predictions. As you probably realize, astronauts in outer space can encounter strange difficulties caused by the absence of gravity. For example, what would you recommend as a bathtub if the astronauts want to take a bath when there is no gravity?

[2]Asymptotes are limiting lines that the branches of the hyperbola approach but never touch.

Three special cases of the contact angle γ.

Clearly, the astronauts should not splash water; for they would have a hard time collecting the floating drops and returning them to their container. Yet a more serious difficulty is that bathtubs may not be able to hold wetting liquids in equilibrium ("wetting" means that the wall of the container exerts "pulling forces" upon the liquid). This is what happens: the free surface of a wetting liquid makes a definite *contact angle* γ (which is somewhere between 0° and 90°) with the walls of its container (see the illustrations on the next page). Suppose that the container has a horizontal bottom and vertical side walls, so that all horizontal cross sections look alike. Assume that the side walls have edges, with the interior angle at an edge being denoted by α. Then the wetting fluid cannot be in equilibrium within the container if $\alpha/2 + \gamma < 90°$. In other words, there is a good chance that the water in the astronaut's bathtub will climb over the upper edges of the container; it will certainly do so if the wetting angle γ is less than 45° and the bathtub has a rectangular cross section.

In many situations, we can assume that the wetting angle γ will be close to zero, say, for glass and a watery liquid. Hence, on your next

A liquid has climbed up the walls of its container (experiment devised by Concus and Finn and carried out by Massica at the NASA drop tower, Cleveland, Ohio); $\alpha = 60°$, $\gamma = 25°$.

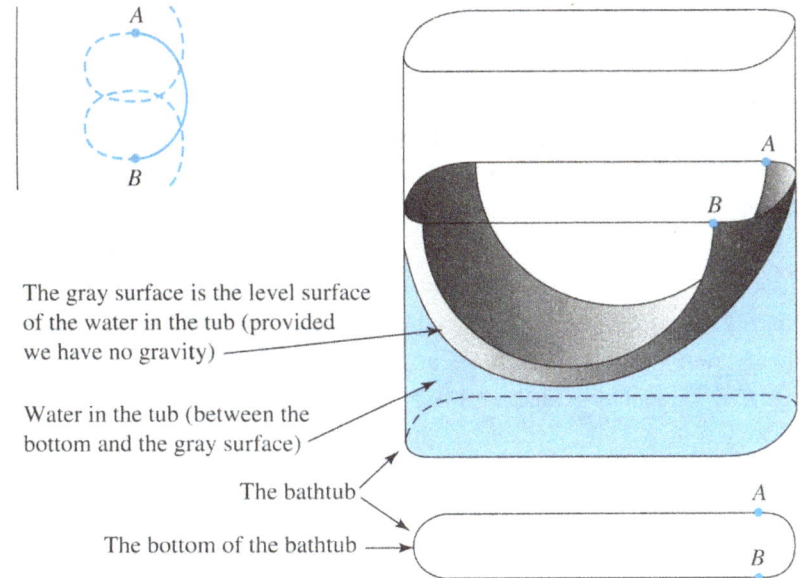

The gray surface is the level surface of the water in the tub (provided we have no gravity)

Water in the tub (between the bottom and the gray surface)

The bathtub

The bottom of the bathtub

Astronaut's bath tub: the roulade of a hyperbola (left) is the curve to which water will cling (right).

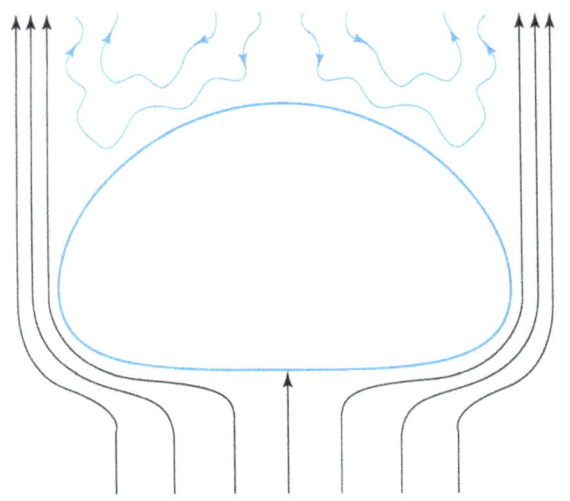

Falling raindrop, and stream lines of air (relative to the drop).

flight to the Moon, you will ruin your clothes if you try to have your aperitif in a glass with a square cross section. This, by the way, is not a mathematical fairy tale, because the formula $\alpha/2 + \gamma < 90°$, inferred from theoretical considerations, was well confirmed by experiments carried out at the NASA drop tower.

The illustration at the bottom of page 239 shows the vertical section of the "Astronaut's Bathtub" suggested by Paul Concus and Robert Finn in 1974. The curvy part of the boundary of this section consists of two pieces of the roulade generated by the focus of a hyperbola. The free surface formed by the liquid in the bathtub is also displayed, assuming that the water meets the boundary walls at an angle of 0°; the water level is part of a nodoid.

There is, however, a third way to get rid of gravity: place the fluid to be investigated (say, oil) into another liquid (alcohol and water in the right mixture) that will not mix with the first one but has the same density. We can, for example, introduce a small amount of oil by means of a tube or a pipette into the alcohol–water mixture. The oil will immediately form a perfect ball.

Every falling raindrop should, of course, be spherical, as it would be if the drop were falling in a vacuum. The air resistance, however, causes the falling drop to be nonspherical. Moreover, there is an internal motion in each falling drop.

Gravity strongly influences drops. For example, on a glass plate, drops of a nonwetting liquid, like mercury, are nearly spherical if they are very small, whereas larger blobs are fairly flattened by their own weight, and large amounts of mercury flow apart and form a very flat puddle. (A nonwetting liquid is repulsed by its glass support.)

Pendant water drops.

The *hanging* or *pendant drops* of a wetting liquid (see the adjacent drawing) are in some ways quite different from *sitting drops*. We all know that a pendant drop cannot be "too large." That is, the elastic

skin can hold the weight of only small amounts of liquid, whereas a heavier weight will tear the skin. There is, in fact, a well-defined size to which a pendant drop can grow. When this size is reached, a tiny perturbation will cause the drop to fall. We can also explain this phenomenon by the saying that the drop has reached its stability limit and has shifted from a stable to an unstable equilibrium. This is why all drops from a medicine dropper are of the same size. It is worth noting that about a hundred years ago numerical investigations of the hanging drop led to the development of powerful numerical techniques that are now widely used by numerical analysts in many fields of applied science.

PLANETS, ROTATING DROPS, AND THE NUCLEI OF ATOMS

Three other phenomena can be explained by a single variational principle founded on Bernoulli's principle of virtual work. These phenomena belong to the fields of astronomy, hydrodynamics, and nuclear physics, which, at first sight, do not seem to have very much in common. Specifically, we will consider *rotating and self-gravitating liquid masses* of homogeneous density, then *rotating liquid drops endowed with surface tension,* and finally the *nuclei of atoms* with or without an angular momentum.

All of these considerations can be subsumed in the following general problem. Imagine some amount of homogeneous fluid, in one or several blobs (or, as mathematicians say, distributed over a connected or a nonconnected domain). Each part of the fluid may be either simply connected (like a ball) or multiply connected (like a solid ring or a solid pretzel).

We shall assume that no exterior forces (such as gravity) are exerted on the liquid, but that all active forces are generated by the fluid itself. This does not exclude the phenomenon of *self-gravitation,* which plays a role only for extremely large masses of a liquid. Self-gravitation is caused by the attraction that every part of the liquid has for every other part according to Newton's law of gravitation. This force tends to keep the liquid together. Let us also assume that

The Moon. Photograph taken by the Lick Observatory, California (36-inch refractor).

surface tension may play a role. In other words, we consider liquids whose surface is covered by an elastic skin that also exerts a contractive influence on the liquid. Moreover, we will assume that two dispersing forces, electrostatic and centrifugal, work against these forces of attraction. If the liquid carries a *uniform electric charge*, then by Coulomb's law each part will repel every other part. If the liquid moves, we even have to take electromagnetic phenomena into account. Finally, if the liquid is *rotating with uniform angular velocity* about its center of mass like a gyroscope[3] (one might think of Jupiter, a huge rotating gyroscope composed of liquid gaseous substances), each part will be subject to centrifugal forces that tend to disperse the liquid.

[3]Whether or not this is a reasonable assumption is another question but as a first approximation it might be acceptable in certain cases.

A rotating liquid body is said to be in *gyrostatic* (or *relative*) *equilibrium* if it rotates like a gyroscope, that is, if there are no "interior motions" within the liquid. An observer sitting somewhere within the liquid and having no connection to the exterior world would not notice the rotation; he would think that the body were at rest. Thus, in gyrostatic equilibrium, four kinds of forces balance each other: the contractive forces of *surface tension* and *self-gravitation* counterbalance the dispersive *electrostatic* and *centrifugal* forces.

The obvious questions to ask are, *what are the possible shapes of liquid bodies in gyrostatic equilibrium,* and *which bodies are in stable gyrostatic equilibrium?*

According to Bernoulli's principle of virtual work, the equilibria are the stationary states of the potential energy, and the stable equilibria

Corona of the Sun photographed during the total eclipse at Green River, Wyoming, June 8, 1918, Hale Observatories, California.

correspond to the minima of potential energy. The total potential energy of a liquid body is the sum of four terms:

*total energy = surface energy + gravitational energy
+ electromagnetic energy + rotational energy,*

where the potential energy of surface tension is in proportion to the surface area of the liquid body and the term "rotational energy" denotes the potential energy of the centrifugal forces.

Historically, the earliest example considered was that of rotating bodies of liquids, which served as models of the planets and, later on, of the stars and nebulae. Here the forces of self-attraction caused by gravitation are so enormously large that the influence of surface tension can be neglected. At first only uncharged celestial bodies were considered; for these the potential energy reduces to the sum of gravitational and rotational energy.

The theory of rotating celestial bodies started with the work of Isaac Newton, Colin MacLaurin, and Alexis-Claude Clairaut. However, experimental observations of the rotation of celestial bodies had, in fact, been made much earlier. The dark spots of the Sun had long been observed without the use of telescopes, and it was known that the spots actually changed their positions. In the European latitudes, the sunspots move from the left-hand side to the right-hand side of the Sun. These observations became much more accurate when astronomers were able to view the Sun with a refracting telescope. In 1611 the German astronomer Fabricius concluded from his observations that the spots were integral parts of the Sun and that their movement was caused by the *Sun's rotation*—a dramatic discovery! In 1612, Galileo in Italy, Thomas Harriot in England, and the German Jesuit Christoph Scheiner published their own observations. Galileo came forth with the same explanation of the movement of the spots as Fabricius, whereas Scheiner said they were small planets revolving around the Sun. In his 1613 publication *Istoria e Dimostrazioni Intorno alle Macchie Solarie e Loro Accidenti*, Galileo disproved Scheiner's reasoning and, for the first time, publicly

supported the heliocentric theory of Copernicus. Scheiner, who eventually conceded that Galileo was right, went on to make much more accurate observations than Galileo had, and he found that the Sun completes a full rotation in 27 days.

The mathematical theory of rotating stars began with Newton's *Principia*, which demonstrated that a slowly rotating liquid body must necessarily be flattened at the poles (the two points that determine the axis of rotation). We saw earlier how the verification of this prediction by Maupertuis finally led to the acceptance of Newtonian physics. Voltaire, who had at first enthusiastically congratulated Maupertuis for his achievements, later wrote the malicious verse

> You have confirmed at places boring and muddy
> What Newton knew without moving from his study.

What could be an equilibrium configuration of a rotating homogeneous liquid body of a given volume V? The simplest spatial configuration other than the sphere would be the *oblate spheroid*. This is a body generated by an ellipse that is revolved about its minor axis, as shown at the right. The spheroid is a special case of a solid *ellipsoid*, i.e. of a body the planar sections of which are all bounded by ellipses (or by circles, which are special cases of the ellipse). Every ellipsoid has *three principal axes*, perpendicular to each other, that pass through the center of the ellipsoid. The *principal radii a, b, and c* are the distances from the center to the end points of the principal axes on the surface of the ellipsoid. Let a be the largest and c the smallest of these three numbers—that is, $a \geq b \geq c$. Then a is the largest and c the smallest distance from the center to the surface. An oblate spheroid is characterized by the relation $a = b > c$, whereas a prolate spheroid satisfies $a > b = c$.

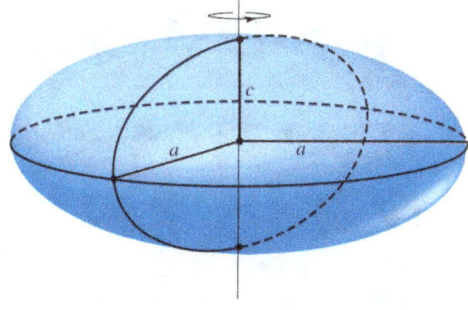

Oblate spheroid, in which the axis of rotation is the minor axis of the ellipse and the major axis is therefore the radius of the circular cross section of the spheroid.

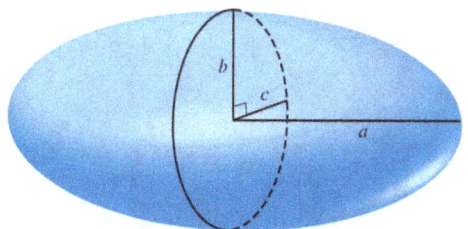

A generic ellipsoid with principal axes and with principal radii a, b, c.

In 1742, when essentially no general theory of fluid equilibrium had been developed, Colin MacLaurin showed the following beautiful result:

> For every volume V and for each angular velocity ω that is not too large, there exist two different rotating oblate spheroids that are in gyrostatic equilibrium.

Here "not too large" means that the angular velocity should not exceed

$$\omega_L = 1.188 \sqrt{GD},$$

where D denotes the density of the liquid and G is Newton's constant of gravitation. If ω tends to the limit ω_L, then the shapes of both MacLaurin spheroids approach that of the same rotating spheroid that rotates with the angular velocity ω_L.

As ω approaches a value of zero (that is, as the rotation slows down to zero), one branch of the MacLaurin spheroids will increasingly resemble a ball of volume V, the well-known equilibrium configuration at absolute rest ($\omega = 0$), whereas the other branch will grow into a disk of "infinite diameter."

For nearly a century it was believed that MacLaurin's spheroids were the only shapes possible for uniformly rotating bodies of homogeneous fluids in gyrostatic equilibrium. Lagrange even claimed that there could not be any other equilibrium configurations; yet this was not true. In 1834, Jacobi discovered that

> For every volume V and every value ω of the angular velocity that is neither zero nor too large, there exists an equilibrium configuration in the shape of an asymmetric ellipsoid ($a > b > c$) that rotates about the axis of the smallest principal radius c.

To be precise, this means that ω should stay below

$$\omega_J = 1.084 \sqrt{GD}$$

if the liquid body has the shape of a truly triaxial ellipsoid ($a > b > c$) that rotates like a gyroscope with the angular velocity ω. Note that ω_J is less than ω_L.

Above: An edge-on spiral galaxy, in COMA BERENICES. Hale Observatories (200-inch).

If ω approaches the value of ω_J, the Jacobi ellipsoid will eventually resemble one of the MacLaurin spheroids (which rotate with the angular

velocity ω_J); if ω approaches a value of zero, the Jacobi ellipsoid will come to resemble a needle of infinite length. What would life be like on a planet that was very thin and very long and that rotated very, very slowly?

Another exciting event was a discovery made in 1885 by Henri Poincaré (1854–1912). He found that a new branch of pear-shaped equilibrium configurations bifurcates (i.e. branches off) from the family of Jacobi ellipsoids, akin to the Jacobi ellipsoids branching off one class of the MacLaurin spheroids. Poincaré conjectured "that the bifurcation of the pear-shaped body leads onward stably and continuously to a planet attended by a satellite." He furthermore proclaimed that along the Jacobi sequence there must be other points of bifurcation that give rise to other stable branches that would eventually develop into planets with two, three, or more satellites. In this way Poincaré envisioned a grand scheme that could explain the birth of our solar system by an evolutionary process rather than by sudden catastrophes.

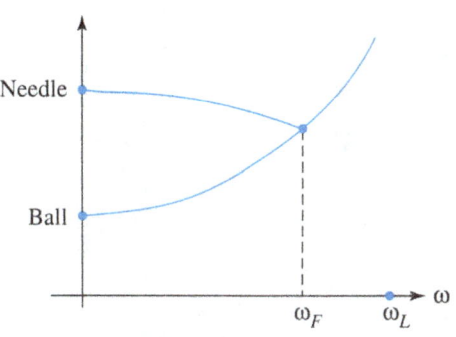

The Jacobi bifurcation diagram.

If we follow the cosmogonic hypotheses of Kant and Laplace, our solar system was at first a huge and slowly rotating gas ball of very low density. Self-gravitation would then lead to a contraction of the gas, thereby increasing density and angular velocity, with the matter then changing from a gaseous into a liquid state. As density and speed increased, the originally sphere-shaped matter would become a more and more oblate MacLaurin spheroid, until the bifurcation point of the Jacobi ellipsoids was reached. At this point, the MacLaurin ellipsoids would lose their stability, whereas the Jacobi ellipsoids would remain stable configurations. For this reason, the liquid body would change into a Jacobi ellipsoid and then, with even stronger contraction, into a pear-shaped body, which eventually would fission into a main body and a satellite.

The Poincaré bifurcation, with pear-shaped body.

 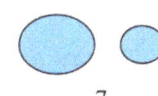

1 2 3 4 5 6 7

A.M. Lyapunov (1857-1918)

Yet Poincaré never made the detailed calculations necessary to substantiate such a scenario. Such calculations were instead carried out by George Darwin (1845–1912), who claimed that he had proved the stability of the pear forms. Unfortunately, Alexander Lyapunov (1857–1918) was able to refute Darwin's calculations, and other scientists reached the same conclusion. Thus Poincaré's wonderful model collapsed. Nevertheless, the theory of equilibrium configurations developed by Poincaré, Lyapunov, and, later, Lichtenstein was the beginning of *bifurcation theory* in nonlinear analysis. This important theory is a principal tool in such diverse areas as fluid mechanics, mathematical biology, and elasticity theory.

To return to our spinning ball of liquid: it was also known that a figure cannot be in gyrostatic equilibrium if its angular velocity ω is too large. There is, in fact, no possible gyrostatic equilibrium if ω^2 is greater than $2\pi GD$. The only way out of this dilemma is to consider rotating liquid bodies in which the liquid is in internal motion, but whose shape does not alter. This is a much "weaker type" of equilibrium, but very likely a more realistic one. It was studied by Dirichlet and Riemann from 1858 to 1860. This work was continued by Chandrasekhar in the 1960s.

Let us now consider small drops of oil that, as devised by Plateau, are suspended in another liquid of the same density. Surface tension is then the dominating force, whereas self-attraction is virtually nil. Moreover, we assume the oil drop to be uncharged. Then the potential energy is the sum of surface and rotational energies (the influence of the friction of the hosting liquid is neglected).

Plateau's apparatus for rotating drops in a liquid.

To follow Plateau's experiments, imagine that a suspended oil drop is touched by a small disk, as shown at the left. This disk is attached to a thin wire and can thereby be rotated, thus causing the drop to rotate by friction. With increasing angular velocity, the influence of the centrifugal forces will become stronger and stronger. First, the oil drop will flatten at the poles and develop a swelling at the equator. At even higher speed, a ring will separate from the drop (see the illustration above) but will rejoin the central drop when the speed dimin-

A rotating drop decomposes first into a drop and ring, then into a drop with droplet satellites.

ishes. However, if the speed increases even more, the ring breaks up into several small drops of *different sizes*.

This circumstance brings to mind a solar system with a large central body circled by smaller satellites. The ring resembles the rings of Saturn or Jupiter. This experiment, however, does not permit us to draw any conclusion about our solar system, because the forces acting in the two situations are rather different.

Recently, Plateau's experiments were repeated and improved upon by scientists at the Jet Propulsion Laboratory in Pasadena. Besides the axisymmetric and ring-shaped figures of Plateau, they discovered two-, three-, and four-lobed equilibrium shapes. With increasing speed, all figures were seen to decay into a one-lobed shape. No satisfactory method of explaining this is currently available because, in fact, the friction between the host liquid and the oil drop cannot be neglected. This friction causes internal flows, which become rather significant as soon as lobes form.

Finally, let us look at the nuclei of atoms. In 1929, George Gamow developed the first model for nuclei, the *drop model*. We can picture a nucleus as a uniformly charged drop with surface tension, either at rest or, in the more subtle model, rotating. The illustration at the top of the next page depicts the potential-energy landscape for a nonrotating nucleus. Here potential energy is shown as "height" above the corresponding states of the nucleus.

Three-lobed figure; arrows show angular velocity of tracers within drop.

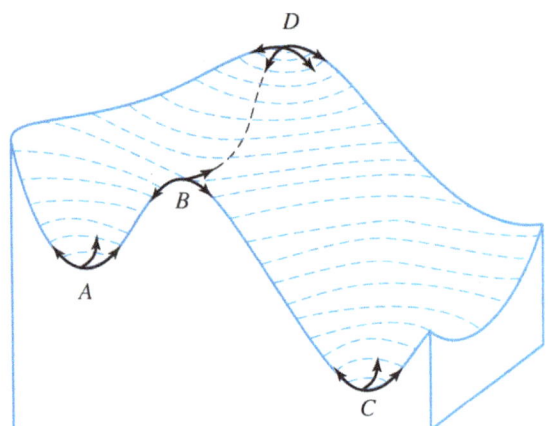

Energy landscape.

The drop will be at the absolute minimum of potential energy if it takes the shape of a ball. If there were no charge, two or any number of balls would also furnish a local minimum of the potential energy. However, if the balls are charged, they will repel each other and will be driven to infinity. Thus several balls "at infinity" (that is, sufficiently far away from each other) are also equilibrium configurations, and the "mountain-pass lemma" tells us that there must be a saddle (mountain pass) in the energy landscape somewhere between the "pits" that represent the one or more charged balls at infinity. This mountain pass represents a physical state of unstable equilibrium.

In an energy landscape that has a single pit corresponding to a spherical configuration separated by a pass from the binary valley of two balls at infinity, the pass corresponds to a symmetric hourglass figure with two equal bulbs. The shape of the hourglass figure depends on the value of a physical parameter (see below). The height of the saddle point above the single-sphere pit is the *energy barrier*

Above: The form of a nucleus at rest in the drop model.

Below: Various forms of a vibrating nucleus.

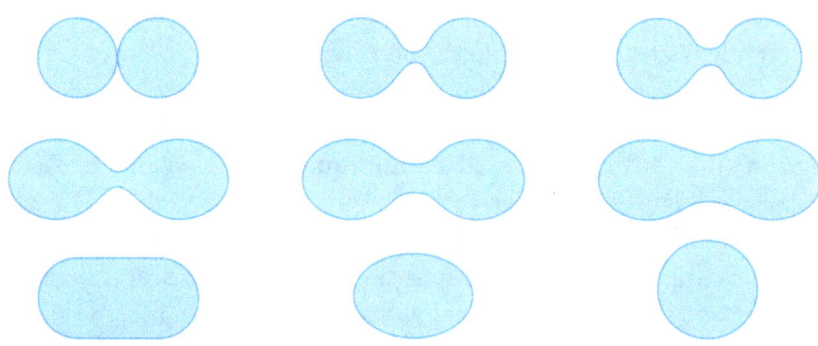

Various hourglass figures corresponding to different values of the physical parameters.

that separates the single-sphere nucleus from fission. If we introduce energy into the nucleus—say, by bumping it with a neutron—we can create vibrations of the nucleus; if these move it above the energy barrier, the nucleus will then be split into two parts.

THE FORM OF LIVING CELLS

Here let us consider the influence and formative power of physical forces on living beings. This is a vast field, full of unexpected marvels, and we cannot do much more than present some fascinating examples and draw your attention to D'Arcy Thompson's *On Growth and Form*.[4] Although this work was never in the mainstream of biological research and, moreover, disregarded genetics and biochemistry alike, it has become a classic in natural philosophy. Biologists have conceded that it has had a great, though "intangible and indirect," influence. Mathematicians, on the other hand, were and still are fascinated by Thompson's considerations.

Radiolarian of genus *Saturnalis*.

Radiolarian of genus *Xiphostylus*.

[4]This book appeared for the first time in 1917. Its enduring fascination has many reasons. One of D'Arcy's colleagues, G. E. Hutchinson, has tried to put it in words: "*On Growth and Form* is a great book because it shows us science as a traditional activity and tells us that tradition is one of daring and of imagination. . . . To foster the values of civilized men takes many talents, and few possess a great share of them. D'Arcy Wentworth Thompson had more than most of us and he used them superbly. What he wrote brings home to the scientific mind, perhaps better than the work of any other writer, what it means to be civilized; the splendor of the panorama that he has presented gives value to what we try to do, and in giving us this he achieved a reward given to a few men."
Another colleague, P. B. Medawar, considered D'Arcy's book "beyond comparison the finest work of literature in all the annals of science that have been recorded in the English tongue."
D'Arcy's father was a classicist who took the education of his son in his hands, and so, in later life, besides his science, D'Arcy was able to translate Aristotle's *Animalium* and write a *Glossary of Greek Birds* and a *Glossary of Greek Fishes*, where he collected and classified all birds and fishes mentioned in the ancient Greek literature, thereby fusing his zoological knowledge with the classics. Medawar wrote, "He was a classicist of sufficient distinction to have become President of the Classical Associations of England and Wales and of Scotland; a mathematician good enough to have had an entirely mathematical paper accepted for publication by the Royal Society; and a naturalist who held important chairs for sixty-four years. . . . In other words, D'Arcy was, according to the best of British tradition, a natural philosopher.

H-surfaces of revolution as unicellulars (compare with the illustration on page 236).

Radiolarian of genus *Tridiktyopus*.

The form, then, of any portion of matter, whether it be living or dead, and the changes of form which are apparent in its movements and its growth, may in all cases alike be described as due to the action of force. In short, the form of an object is a "diagram of forces," in this sense, at least, that from it we can judge of or deduce the forces that are acting or have acted upon it: in this strict and particular sense, it is a diagram—in the case of a solid, of the forces which have been impressed upon it when its conformation was produced, together with those which enable it to retain its conformation; in the case of a liquid (or of a gas), of the forces which are for the moment acting on it to restrain or balance its own inherent mobility. In an organism, great or small, it is not merely the nature of the motions of the living substance which we must interpret in terms of force (according to kinetics), but also the conformation of the organism itself, whose permanence or equilibrium is explained by the interaction or balance of forces, as described in statics.

We should bear in mind two facts: biological phenomena are exceedingly complex, and not everything that seems obvious is true. We should take the mathematical models of these phenomena only as metaphors, and not as "laws" as in astronomy or physics. The day may come, however, when biological forms and processes can be described with similar mathematical precision as those in physics.

Let us consider the influence of surface tension on unicellular beings. They are, crudely speaking, drops of protoplasm—of a very viscous fluid—which are suspended in water. We therefore should expect them to be of spherical shape. This is what we see, in fact, with many simple organisms.

However, as Thompson observed, the spherical forms are usually passive, whereas the unicellular entities that differ greatly from sphericity seem to be rather mobile and quite often possess little whips in the form of flagella or cilia. In Thompson's opinion, these whips tend to support forms that would otherwise be unstable, and they seem to play a role like that of the boundary configurations (frames) in experiments with soap films and soap bubbles. Moreover, Thompson found all types of *H*-surfaces of revolution among unicellulars. These forms, as he noted, would often have

been unstable if it were not for the cilia in permanent motion at the "free edges."

He was cautious enough to state that it is not clear whether *true surface tension* is responsible for all these phenomena or whether *membrane tension* might play a role (unlike a liquid film with surface tension, a membrane contracts just so much, then no more). Modern experimental work has indeed shown that both kinds of tension act in the boundary of a cell. But, as Thomson remarked, within wide limits both forces produce very similar forms.

Thomson also investigated the siliceous skeletons of the radiolarians and made a step toward explaining how these skeletons might be formed. He compared the radiolarian body with a

> . . . frothy protoplasm, bubbled up into a multitude of alveoli or vacuoles [small cavities], filled with a fluid which can scarcely differ much from sea water, . . . According to their surface-tension conditions, these vacuoles may appear more or less isolated and spherical, or joined together in a "froth" of polyhedral cells; and in the latter, which is the commoner condition, the cells tend to be of equal size, and the resulting polygonal meshwork beautifully regular.

(A bubble froth has to satisfy the two rules, found experimentally by Plateau, that were described earlier in Chapter 5.)

Radiolarian of genus *Cyrtocalpis*.

Thompson then suggested that adsorption forces were particularly strong in the edges of the froth, which therefore lead to inorganic materials being deposited along these edges, thus creating the siliceous framework of the radiolarian skeleton. Hence *the marvelous regularity of the skeleton was nothing but the materialization of Plateau's two rules*. This suggestion, however, does not explain why there are so many different kinds of skeleton forms among the many species of radiolarians. We are still as puzzled as Ernst Haeckel must have been when he made 4,700 drawings of the radiolarians that he had studied under his microscope. Yet these little "orchids of the sea," as they have been called, are surely no joke of evolution. On the contrary, since radiolarians are among the oldest living creatures on the

Radiolarian of genus *Cecryphalium*.

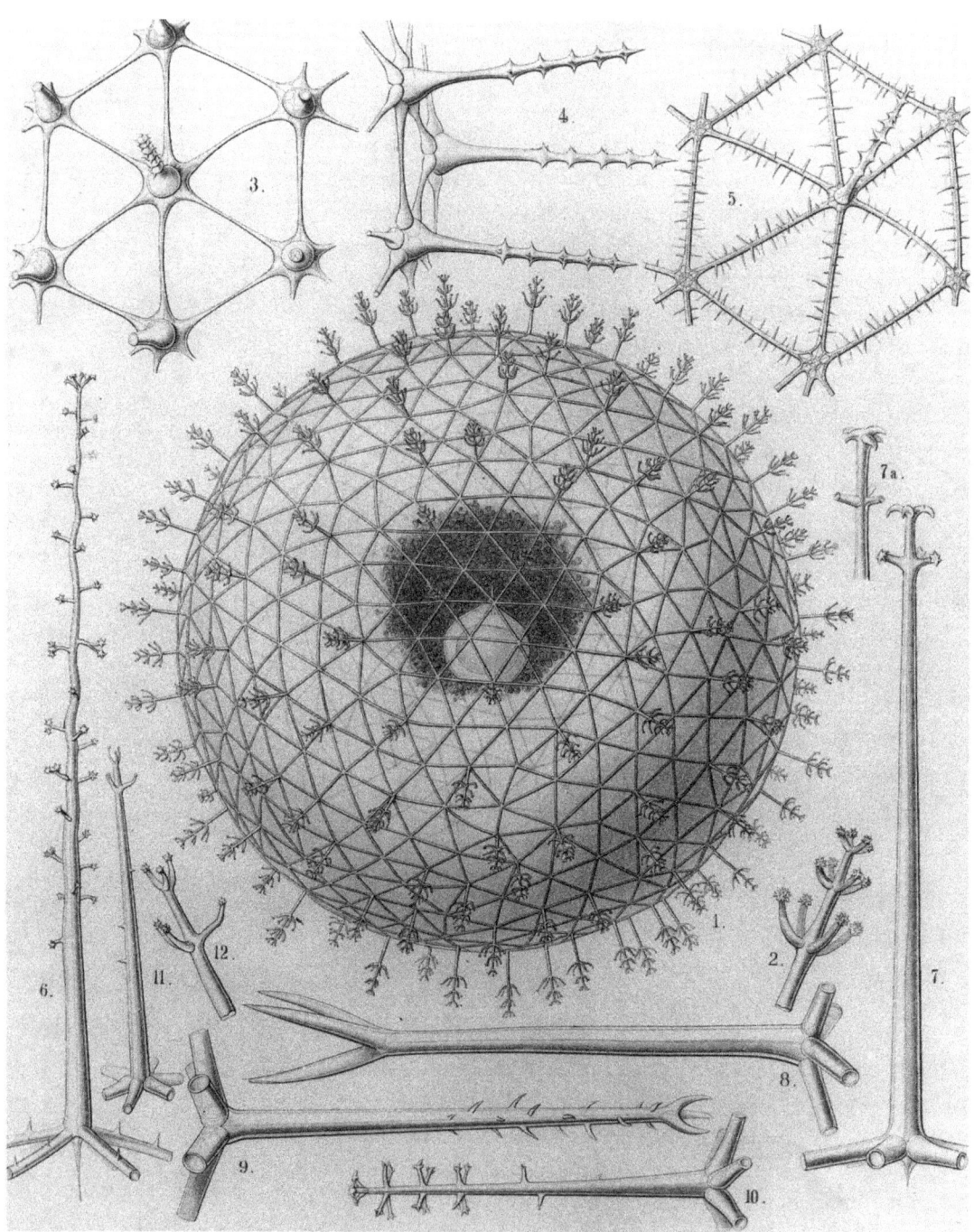

A radiolarian of genus *Aulosphaera*, drawn by Ernst Haeckel (*Report on the Radiolaria collected by H.M.S. Challenger during the Years 1873–1876*).

Radiolaria of the genera *Stylochromium, Caryostylos, Staurolonche, Staurocaryum,* and *Rhizophlegma*, drawn by E. Haeckel (*Report on the Radiolaria collected by H.M.S. Challenger during the Years 1873–1876*).

A drawing of a radiolarian by E. Haeckel.

Earth, they are a kind of living fossil, and their skeletons must be very sound constructions. Unfortunately, we have no idea why nature arrived at these forms, although experience has taught us to expect that, for the radiolarians, such forms must be in some way beneficial.

Recent investigations of living radiolarians under an electron microscope have brought forth another astounding fact about their geometry. Each radiolarian extends many (sometimes hundreds) of very thin, elastic, and thoroughly straight threads (called *axopodes*), which are necessary for metabolism, through the tiny holes of the skeleton. The transportation of matter is carried out by the cell through extremely narrow tubes called *microtubuli*. Now, amazingly enough, these microtubuli are very regular. For example, some cross sections have shown a perfectly hexagonal array, like the honeycomb pattern. The hexagonally ordered microtubuli are connected by bridges that function as pumps.

Electron micrographs of two skeletons of radiolaria, prepared by Manfred Kage.

Thus we see that nature seems to prefer order and regularity even in its tiniest structures. Today it seems to have been proved that tension forces at the surface of a cell only partially explain its shape; it is very likely that internal structures, such as microtubuli, are to a large extent responsible for cell shape.

CRACKS AND FISSURES

Fractures and fracture patterns arise often in nature. Mountain climbers dislike crevasses particularly when they are hidden under snow cover. There are cracks in the bark of an old tree or in the wooden beams that support the roof of a house. Of particular interest are polygonal fracture patterns, which may be seen in dried mud, in the varnish of old paintings, in ceramic glazes, and in enamel. We also find them in concrete, basalt, and limestone, in arid regions, and in frozen earth. Some patterns are formed of tiny tiles; others are

A.

B.

C.

A. Cracks with a 120°–pattern in the salt crust forming a hexagonal pattern of cells (of about one yard in diameter). Devil's Golf Course, Death Valley.
B. Tesselation of arid land by sun cracks in Nevada.
C. Mud cracks in a container filled with wet soil.

Crack formation by shrinking of plaque on a tooth.

made up of gigantic polygons. In the playas, the dry lakes of the Big Basin area of the western United States, aerial photographs of polygons between 100 and 300 feet wide have been made. The ground-ice wedges in Alaska and the Tundra polygons are well-known and often-described phenomena seen on the permafrost in polar regions.

Most scientists currently agree that the polygonal patterns seem to appear in materials that undergo volume changes because of drying or cooling, or both. The more cautious among them admit that we do not understand these processes well at all. Others see Bernoulli's principle of virtual work in action; yet the conclusions drawn from this principle differ from scientist to scientist. Some claim that the cracks have to intersect each other perpendicularly; others, that crack polygons must be pentagonal or hexagonal. In particular, the hexagon pattern, with the three-rayed vertices and the three 120° angles, is supposed to result from this mechanical principle. Not every plane domain can be covered by hexagonal tiles, because Euler's formula imposes a restriction on potential tessellation (see page 174); so we usually find a few pentagons among many hexagons. Similarly,

skeletons of radiolarians with mainly hexagonal cells are never purely hexagonal; we easily spot a few pentagons and heptagons.

We think that these "theoretical" conclusions are neither convincing nor well established, although they probably contain some truth. The theory of tearing and splitting of strongly viscous or brittle materials is not well developed. Moreover, we know that the most minute irregularities in a material might obviate all theoretical calculations. For example, it is rather difficult to tear a strip of paper with straight edges by pulling it at both ends, but a tiny notch at one edge makes the tearing easy.

It is well known that irregularities appear in most substances; hence it seems safe to assume that irregularities play an important role in the formation of cracks. The majority of earth fissures seem randomly distributed over the ground, and a new crack in a formerly uncracked area will develop in randomly formed zones of weakness in the ground. Moreover, fissures will not usually all appear at the same time. We find that at first only a few cracks develop. The tangential stress along the fissures in the crack zones is still quite formidable, whereas the perpendicular stress is released by the formation of the cracks. Therefore, secondary cracks will form perpendicularly to the greatest tension and thus tend to intersect the primary cracks at right angles. The ternary cracks will then meet the primary and secondary ones at right angles. In this way, a network of randomly distributed fissures that intersect perpendicularly will form. The history of the fractional pattern is therefore well documented. We can usually decide which fissure came earlier and which came later.

Pentagons and hexagons in a radiolarian skeleton.

Unfortunately, the general situation is not as clear-cut. First, there are materials in which the cracks form strictly 120° angles. (A close inspection shows much narrower secondary fissures that meet the primary ones at a right angle.) Second, careful investigations have shown that angles of 90°, 120°, and 60° can appear. For example, of a large number of crack intersections measured at islands in the Mackenzie Delta, 80 percent were perpendicular; the 60° angle was

Incipient postcrack pattern on Kendall Island, NWT, Canada.

the next most common, but very few triradial intersections with 120° angles were found (see illustration above).

There is a way to explain the 60° angles in terms of the principle of virtual work. Suppose that we assume that the frozen ground started to form a triradial array with 120° angles. Then unevenness of the frozen soil might have prevented the formation of the proper third ray. If this were a sufficient explanation, the mechanical principle of crack formation could be saved. Accordingly, cracks would be formed either as single cracks (perpendicular to the direction of greatest tension), or as triple cracks with the 120° angles in a homogeneous material (yielding maximum release of stress with minimal strain), or else as the hybrid form, with the 60° angle caused by the "notch effect." Later cracks would be generated accordingly, but they would intersect former ones at right angles. We suggest that readers carefully observe natural cracks and formulate their own theories about them.

Giant's Causeway, Ireland. A vulcanic rock formation, consisting of basalt pillars whose cross sections are more or less regular hexagons.

THE GEOMETRY OF CRYSTALS

Among the most-admired forms in nature are those of crystalline structures. Crystals, often radiating deep, rich colors, are as much sought after by collectors as studied by scientists. A natural question to ask is whether the shapes of crystals can be explained by a variational principle. It is known that the form of a growing crystal is determined by several kinetic factors, such as the mechanism of transport of material, condensation, diffusion, and rates of chemical reactions. A complete understanding of surface structures is consequently difficult.

A calcite twinned crystal from the Elmwood Mine in Tennessee.

Crystals. (A) Hessonite, a kind of Grossularite. (These crystals belong to the class of garnets). (B) Pyrite. (C) Spinel. (D) Anatase. (E) Fluorite. (F) Boracite.

However, for small crystals or for crystals with small surface irregularities, it appears that a crystal's tendency to lower its free surface energy is the dominating factor in its shape formation. We are therefore led to ask what shape a small crystal of *fixed* volume will assume if its free surface energy is to be at a minimum. This reminds us of the isoperimetric problem in three dimensions, the solution of which is a sphere. In this case, the potential energy is proportional to the surface area, whereas the volume is fixed. The perfectly smooth and symmetric sphere is clearly quite different from crystalline structures. Nevertheless, when crystals are small, their shapes can be explained to some extent by a variational principle similar to that of the isoperimetric problem, and the remarkable difference in structure results from the difference in the corresponding potential energies.

The mathematical problem can be formulated as follows. Consider a region in space bounded by a finite collection of pieces of smooth surfaces. Such bodies are called *piecewise smooth*. For almost every point on such a piecewise smooth surface, there will be a tangent plane. The surface energy for crystalline structures depends on the chemical nature of the specific crystal, and we shall assume that this energy depends only on the orientation of the tangent planes in space.

We then ask for the structure or shape that, for a fixed volume, has a minimum *total* surface energy. The noted discovery made by the crys-

A piecewise smooth body.

tallographer Georg Wulff in 1901 is that, given some further reasonable assumptions about the mathematical character of the surface energy, the following holds:

> For every given volume, there is a unique convex body whose boundary consists of planar faces, such that this boundary surface has less energy than does the boundary surface of any other piecewise smooth body of the same volume.

This theorem is remarkable in two ways. First, there is an infinite number of possible surface energies; nevertheless, for each such admissible energy, the unique minimum is a convex region bounded by planes. Second, unlike most problems in mathematics in which explicit solutions are impossible to find, the solution to our minimum problem, the optimal crystalline region, can be determined by a simple procedure known as the Wulff construction.

Wulff's construction is based on two assumptions. First,

> the surface energy per unit area at any point on the surface depends only on the direction of the tangent plane to the surface at that point.

To each plane in space we can associate a "vector" (or arrow) of length 1 which is perpendicular to the plane. Such an arrow indicates the

A

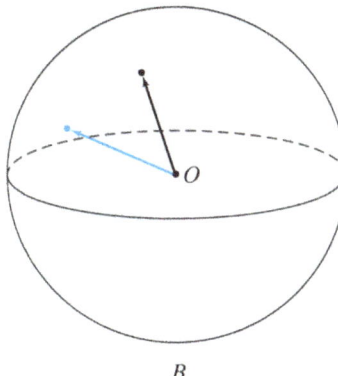

B

The directions normal to all planes form a sphere.

orientation of the plane in space, as shown in part A of the illustration on the preceding page. We now fix a point O in space, and then without changing each arrow's direction we can transport them so that their tails are fixed to O. Thus each arrow is completely determined by its tip. The set of all such tips fills out a sphere in space, as shown in part B. By this correspondence, we can think of the surface energy as a rule that assigns a positive number (the energy per unit area of a piecewise smooth body) to every point x on the sphere.

The second hypothesis on which Wulff's construction is based is:

> For a given crystal, there are N distinct points p_1, p_2, \ldots, p_N on the sphere of perpendicular directions and a "triangulation" of this sphere with these points as the vertices of the spherical triangles. The value $E(x)$ of the energy at any point x in the interior of a spherical triangle satisfies a certain inequality with respect to the vertices of the triangle in which x sits. If, for simplicity, we say that p_1, p_2, and p_3 are the vertices of such a triangle, then, for some positive numbers a_1, a_2, a_3,
>
> $$x = a_1 p_1 + a_2 p_2 + a_3 p_3.$$

A triangulation of the sphere.
(See also page 254.)

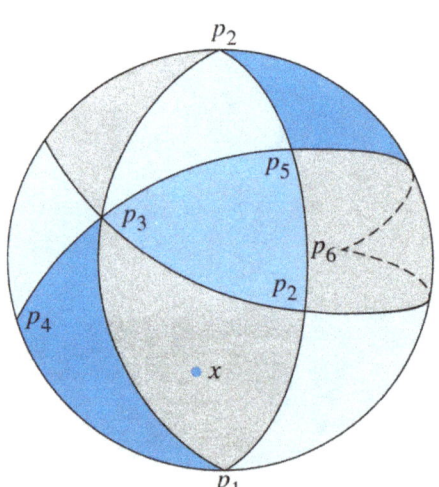

The assumption is that

$$E(x) > a_1 E(p_1) + a_2 E(p_2) + a_3 E(p_3).$$

If x lies on the edge of such a triangle, a similar inequality is assumed to hold. Because the energy E assigns positive values to a point on the sphere, we can think of plotting E as a "graph" over the sphere.

We obtain this graph as follows. For every point z on the sphere, we move outward, along the radial line through z, for a distance of $E(z)$ from z. Such graphs are difficult to visualize. A cross section, depicting what such a graph looks like in two dimensions, can be obtained by having the graph over the sphere intersect a plane, as shown at the right. We may now proceed with the construction of the crystalline shape with the minimum possible total surface energy. For each vertex point p_i in the sphere, we proceed outward from the center of the sphere along the radial line through p_i until we reach a point at a distance $E(p_i)$ from p_i. Then we draw the plane perpendicular to this radial line at this new point. The optimal shape is the bounded convex set determined by all these planes.

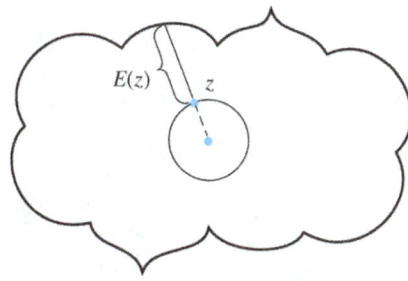

A cross section of the graph of a crystalline energy function.

A planar cross section of such a shape is depicted at the right. This shape, however, may not have the required volume. The crystalline body that solves the problem is obtained by rescaling this shape so that it encloses the desired volume. Wulff did not have a complete proof of this result. In 1914, Heinrich Liebmann showed that if there is an equilibrium shape for the crystal, it must be the polyhedron obtained from the Wulff construction; so the question of existence was again left open. Finally, in 1944, Alexander Dinghas proved that any shape differing from that obtained by the Wulff construction has a higher total surface energy. Thus the beautiful and jagged crystalline form that we see in quartz and diamonds can, in some sense, be explained by minimum principles, as can the spherical and spheroidal forms that we see in planets, soap bubbles, and oil globules.

The Wulff construction.

Isaac Newton (1642-1727). Painting by I. Vanderbank (1725).
London, National Portrait Gallery.

Epilogue

Dynamics and Motion

In the preceding chapters we have discussed both the application of variational principles in various physical situations and their influence on the development of mathematical analysis. We have, until now, mainly restricted our attention to physical systems *at rest*, that is, in equilibrium; but the description of *motion* is equally important.

We know that the Greeks had wondered about the shape of the trajectory of a stone thrown into the air and of the orbits of the planets. Yet the emphasis of Greek science was more on the geometric shape of the celestial paths and less on the *motion in time* along these orbits. If, as in astronomy, such a motion was of special interest, it was explained as a uniform motion or, more intricately, by a combination of uniform motions, as in the epicycle theory, which fitted the astronomical data obtained from the observations.

It is important to realize that the ancient Greek philosophers saw no relation between, for example, the motion of the planets and that of a falling stone. Such a connection was not drawn until the seventeenth century, the age of Kepler, Galileo, Huygens, and Newton. For the Greeks, the planets were attached to the celestial spheres, and no explanation was given as to what kept them moving. Concerning earthly bodies, Aristotle said that objects tend toward their natural place, which for heavy objects was the center of the Earth, but for smoke it was in the sky. It is not surprising that the Greeks never developed a precise kinematics, that is, a description of all possible motions in time, since they failed to invent a tool similar to infinitesimal calculus. Certainly, Archimedes could compute some areas and volumes, using ideas close to those of integral calculus, but the concepts used in differential calculus were never developed in antiquity. Therefore, the Greeks did not arrive at a precise description of changes in motion; for example, they had no clear understanding of the notion of *velocity*, not to mention *acceleration*. Only the discovery of infinitesimal calculus by Newton and Leibniz and its elaboration by the Bernoullis and Euler provided scientists with the decisive mathematical tool enabling them to form a theory of physical motion.

The foundations of such a theory were laid in 1687 by Isaac Newton with the publication of his treatise *Philosophiae Naturalis Principia Mathematica* (Mathematical Principles of Natural Philosophy), possibly the most important and certainly one of the most influential scientific books ever written. In this masterpiece, Newton explained all possible motions of physical bodies in terms of the influence of *forces*

acting upon them; consequently, this theory is called *dynamics* (from the Greek word *dynamis*, meaning "power" or "force").

Looking into Newton's treatise, we might expect to find dynamics to be developed by means of infinitesimal calculus (or, in Newton's terminology, by *Analysis per Quantitatum Series, Fluxiones ac Differentias*). However, little of this sort is true; in fact, not much analysis by means of fluxions (i.e., by differentiation) is visible at all. On the contrary, Newton formulated all of his results in the style of Greek mathematics, presenting every proof in a geometric–synthetic manner. Nevertheless, the mathematical structure of the *Principia* is entirely nonclassical as it is built by means of "limit operations" for fractions of varying quantities, unknown in antiquity, combined with basic material from Euclidean geometry. Only very rarely did he appeal to some earlier work on infinite series and fluxions that he had written in 1671. However, when the priority quarrel with Leibniz over the discovery of calculus erupted, Newton wanted to prove that he, from the very beginning, had applied his calculus to problems in physics and that moreover all the propositions in the *Principia* were obtained by means of his analytical methods developed in the *Quantitatum Series*. He wrote,

> The Propositions... were intended by Analysis: but... I composed what I invented... to make it Geometrically authentic & fit for the publick... But if any man who understands Analysis, will reduce the Demonstrations of the Propositions from their composition back into Analysis... he will see by what method of Analysis the Propositions were invented.

His claim of priority infiltrated our scientific literature, but we do not know exactly how substantiated it is. Derek Thomas Whiteside, the editor of Newton's *Mathematical Papers*, stated to have been unable to find any manuscript preceding the *Principia* which would prove that Newton had *first* obtained the mathematical results of the *Principia* by means of his fluxions, before transforming them into the traditional geometric form found in the *Principia*. This does not contradict the fact that the *spirit* of the *Principia* is that of infinitesimal calculus. In

Frontispiece of Newton's notes on Analysis published by W. Jones in 1711.

Euler's *Mechanica* (1736).

Frontispiece of Euler's *Theoria Motus* (1765).

fact, Michael Nauenberg recently contested Whiteside's statement by pointing out that there is enough evidence showing that, at the time of the *Principia* and even before, Newton used calculus to solve problems in mechanics.

The first text formally basing mechanics on differential and integral calculus was Euler's 1736 treatise *Mechanica sive Motus Scientia Analytice*. This brilliant reformulation of Newton's mechanics was supplemented about thirty years later by Euler's second book on mechanics, the *Theoria Motus* (1765). A century after Newton's *Principia*, Lagrange published his celebrated *Méchanique Analitique* (1788). Here the basic law of mechanics is expressed in the form of a disguised *variational principle*, namely, as a version of the principle of least action, which Lagrange had formulated as an analogue of Johann Bernoulli's principle of virtual work (see Chapter 4). In a somewhat different form, this principle had already been stated by Jean le Rond d' Alembert (1743). Lagrange's action principle can be viewed as a true extension of Newton's original point mechanics since, for example, it allows one to study the motion of systems of rigid bodies.

One of the interesting features in Lagrange's approach to mechanics is his interpretation of time as another dimension, equivalent to the three spatial dimensions. At first this point of view was not much more than a mathematical artifice, but Lagrange's idea, together with some remarkable physical experiments, eventually led to the theory of relativity. Physicists and mathematicians had become accustomed to a four-dimensional world, and in 1908, Hermann Minkowski began his lecture on "Space and Time" with the following words:

> From this hour, space and time taken for themselves shall be mere shadows, and only a kind of union of both shall still keep autonomy.

From Minkowski's point of view, dynamics is geometry in a four-dimensional world, just as statics is geometry in our old three-dimensional world. Thus the belief of the ancient Greeks that everything in our physical world reduces to geometry is in some sense realized. Einstein's basic *field equations*, which replace the old laws of Newton,

Hermann Minkowski (1864–1909).

follow from an "action principle" that, in fact, is a purely geometric variational principle determining the shape of Minkowski's four-dimensional space.

We now consider some of the ideas that led first to Newton's mechanics and then to Einstein's theory of relativity, and we shall then look at their connection with variational principles.

DYNAMICS BEFORE NEWTON, AND THE SOURCES OF THE *PRINCIPIA*

In the first half of the seventeenth century, a new epoch in mechanics began when Kepler and Galileo stated precise laws describing how a planet moves with nonuniform speed along its elliptic orbit, or how a stone falls to the earth on its parabolic trajectory. At about the same time, Descartes conceived a new "philosophy of nature." He proposed to explain all physical phenomena as consequences of *conservation laws* and *collision laws*. The Cartesian model of the world dominated French science for nearly a century until it was replaced by Newton's theory.

The laws of collision were, after the first attempts by Descartes, studied by the English mathematician John Wallis (1616–1703), by Christopher Wren (1632–1723, the architect of St. Paul's Cathedral in London), and in particular by Christiaan Huygens, who achieved the most remarkable results. Using the *law of conservation of total energy*, Huygens successfully treated elastic collisions as well as the oscillations of the physical pendulum. He avoided using the notion of force in these investigations, whereas this concept played a prominent role in the thoughts of Galileo, who wrote,

> If no force acts on a body and the body is at rest, it will remain at rest; if no force acts and the body is in motion it will continue to move at constant speed in a straight line.

In the middle of the seventeenth century, it was fully understood that *a change of velocity requires forces*. But the relation between the change of velocity and the acting force was not yet clear. A careful

Frontispiece of the Minkowski paper from 1909.

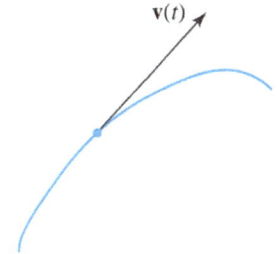

The velocity vector $v(t)$ of a car moving on a road is an arrow whose length equals the speed of the car at time t, while its direction tells which way the car is moving.

study of the change of speed for curved motions was needed. If, for instance, an observer moves with constant speed along a straight line, his "velocity vector" does not change, and the observer does not experience any forces different from those in rest. On the other hand, if he moves with constant speed along a circular path, as on a merry-go-round, his velocity vector is always changing and he feels the influence of a *centrifugal force*. Descartes had already stated that an ant sitting on a turning stick must hold fast and that a stone in a rotating tube tends to move away from the center of motion.

In 1665, Huygens investigated the centrifugal force F acting on a point-mass that moves with constant speed v along a circle of radius r. He found the correct relationship between these quantities, namely that F is in proportion to v^2/r:

$$F \sim v^2/r.$$

This relation can be written in a slightly different form if we introduce the *angular velocity* ω, the rate of change of the center angle of the turning wheel. We find that the speed v is given by the formula $v = \omega r$, and thus the centrifugal force is proportional to $\omega^2 r$:

$$F \sim \omega^2 r$$

Huygens published his results in the *Horologium Oscillatorium*, his 1673 treatise on the pendulum clock that we mentioned in Chapter 3.

In 1644 the French scientist de Roberval had postulated that all material bodies exert an attractive force on each other, and he explained the attraction by the Sun as a special case of such forces. Moreover, in a lecture given to members of the newly founded Royal Society on May 23, 1666, the English physicist Robert Hooke (1635–1703) interpreted the curvature of the planetary orbits around the Sun as a consequence of the Sun's attraction, and he supported his theory by pendulum experiments. In 1674, Hooke formulated a dynamics of celestial motions, and he assumed the existence of an attractive force exerted by the Sun that decreased as the distance from the Sun increased. This theory, however, was only qualitative, and Hooke

could only speculate how the attraction between the Sun and one of its planets depended on their distance. Hooke did carry out experiments to measure how the strength of the Earth's attraction on a stone depended on the height of the stone above the ground; but even by climbing the highest towers of London, he could not find any measurable effects from which an answer could have been inferred. The differences in the strength of attraction are so small that he was unable to establish them with the techniques of his time.

During the period of 1679–1680, Hooke discussed his ideas about gravitation with Newton; this correspondence is preserved and has by this time been published. One knows from Hooke's diary that he had asked some mathematicians if they could use his physical principles to compute the motion of planets, but he did not find any help. So he turned to Newton asking for his advice. On November 24, 1679, he wrote to Newton, "But my supposition is that the Attraction always is in a duplicate proportion to the distance from the Center Reciprocall." This is the first time Hooke stated that he believed the force of gravity, F, depends inversely on the square of the radial distance, r, that is,

$$F \sim 1/r^2.$$

One is led to this form of the gravitational force by a simple reasoning. Let us assume that a planet moves uniformly around the Sun along a circle of radius r (which is the special case of an elliptic orbit), with an angular velocity ω. Then Kepler's third law (see Chapter 2) states that $\omega^2 r^3 = K$, where the number K is the same for all planets, and so we have

$$\omega^2 \sim 1/r^3.$$

On the other hand, if the planet is to move uniformly on a circular orbit, we know that the centrifugal force $F \sim \omega^2 r$ must be counterbalanced by a centripetal force of the same size which is directed toward the Sun. Therefore, the attractive force of the Sun has to be in proportion to $1/r^2$.

This is more or less the reasoning that led Hooke, as well as Wren and Halley, to the *inverse-square law of gravitation*. However, they were unable to deduce the three Keplerian laws from this form of the gravitational force. In particular, this required one to show that a planet attracted by the Sun according to the $1/r^2$-law moves along an elliptical orbit, with the Sun at one of its foci.

On January 17, 1680, Hooke wrote to Newton, "I doubt not but that by your excellent method you will easily find what that Curve [i.e., the orbit of the planet attracted by the $1/r^2$-law] must be, and its proprietys, and suggest a physicall Reason of this proportion . . ."

Newton eventually solved the problem but he never answered that letter from Hooke, and he later denied having learned anything from Hooke, admitting only that Hooke's letters had stimulated him to reinvestigate the basic problems of celestial mechanics. Instead, he insinuated that he had solved the riddle of gravity rather early in his youth, and thus the legendary falling apple might be traced to these later accounts of Newton.*

After the correspondence with Hooke, Newton's interest in celestial mechanics temporarily subsided. It was later rekindled when Edmund Halley (1656–1743) communicated to him Wren's challenge to determine, theoretically, the paths of the planets. It is not certain at exactly which time Newton had solved the problem, but when, in August 1684, Halley visited Newton in Cambridge and asked him what the orbit of a planet under the influence of an inverse-square force would be, he answered that it had to be an ellipse. As the stunned Halley inquired how he knew this, Newton's reported reply was, "Why, I have calculated it." Halley requested that Newton publish his results. Newton then wrote a paper, "Propositiones de Motu" (Results on Motion), in which he described his main results about the motion of bodies attracted by a central force. When the

Edmond Halley (1656–1743) discovered that comets may move periodically, and several historical comets were explained by him as the reappearances of one-and-the-same comets. For example, he predicted the appearance of the comet from 1680/81 ("Halley's comet"), which was identical with the comets from 1456, 1531, and 1607.

*However, Michael Nauenberg maintains that Newton used calculus to discuss the main propositions of the *Principia*, and he states: "Newton's claim that *"by the inverse method of fluxions I found in the year 1677 the demonstration ... that the planets move in ellipses ..."* can simply not be dismissed as bogus history as is the view of many Newtonian scholars today." (See *Archive for History of Exact Sciences*, vol. 46, 1993-94, pp. 221-256, and in particular p. 249.)

Royal Society received this work, Hooke became quite upset and claimed that the principal ideas were his. Newton understandably fumed; his feelings are expressed in a letter to Halley (June 20, 1686): ". . . is this not very fine? Mathematicians that find out, settle and do all the business must content themselves with being nothing but dry calculators and drudges and another who does nothing but pretend and grasp at all things must carry away all the invention. . . ."

Halley urged Newton to turn his results into a book, and he unselfishly supported Newton in many ways. Newton expanded "De Motu" into the *Principia*, and on April 28, 1686, the manuscript was submitted to

PHILOSOPHIÆ NATURALIS PRINCIPIA MATHEMATICA.

Autore *JS. NEWTON, Trin. Coll. Cantab. Soc.* Matheseos Professore *Lucasiano,* & Societatis Regalis Sodali.

IMPRIMATUR·
S. PEPYS, *Reg. Soc.* PRÆSES.
Julii 5. 1686.

LONDINI,

Jussu *Societatis Regiæ* ac Typis *Josephi Streater.* Prostat apud plures Bibliopolas. *Anno* MDCLXXXVII.

The frontispiece of the two-lines print of the *Principia* carrying the imprint "Prostat apud plures Bibliopolas," which is sometimes called the "first issue" of the first edition, while the "export copy" (with the three lines "Prostant Venales apud Sam Smith . . . aliosq; nonnullos Bibliopolas") is called the second issue of the first edition. This distinction between first and second issue seems to be quite unfounded. It has been suggested that Halley made an agreement with Smith concerning foreign sales; in fact, most of Smith's fifty copies were apparently sold on the Continent.

the Royal Society. The council of the Society decided on May 19 that "Mr. Newton's work should be published forthwith." Since the Society's treasury was without funds because of a prior expenditure for the publication of a book on the habits of fish, Halley paid for the publication fees himself. Thus, in 1687, the first edition of Newton's Principia appeared, a work that Laplace, himself a great scientist, later called "the most outstanding production of human genius of all time." It is unknown how large this first edition of the *Principia* was; the estimates range between 250 and 400 copies, and about 200 copies are still known to exist.

The frontispiece of a special copy of the *Principia*. This is the celebrated copy with marginal notes by Leibniz, which was rediscovered by the Swiss historian of science E. A. Fellmann in 1969, and is kept in the Bibliotheca Bodmeriana at Cologny, near Geneva, Switzerland. The figure shows the three-line frontispiece of the "export copy."

Apparently, the book was well received in England. It was read on the Continent by the leading mathematicians and astronomers but was not as well accepted, since the Cartesian philosophy still dominated continental Europe. Huygens, himself a Cartesian, thought the idea of gravitation to be absurd and wondered why Newton had made so many calculations based on absurdities. As we have seen, it took about fifty years before the power of the "Newtonian philosophy" was fully recognized throughout Europe. The unfortunate priority quarrel between Newtonians and Leibnizians over the invention of infinitesimal calculus may have played a role in this delay. In England, however, Newton soon became famous, and Newton's epitaph by Alexander Pope,

> Nature and nature's laws lay hid in night; God said, "Let Newton be!" and all was light.

succinctly describes what an impression Newton must have made upon his English contemporaries.

THE DYNAMICS OF NEWTON

The *Principia* consists of three books, the first of which deals with the motion of one or more bodies in empty space. In addition, numerous problems connected with the law of the inverse square, such as the attraction of a spherical shell and of solids of revolution, are treated. The motion of the pendulum is discussed, and, at the very end, the laws of reflection and refraction of "light corpuscles" in media of varying density are treated as mechanical problems. The problem of how a body is attracted to a fixed center according to the inverse-square law is completely solved, and Newton shows that the problem of how two bodies move if they are attracted to each other by the inverse-square law can be reduced to the first one. The "three-body problem" (i.e., the Sun and two planets) is studied, but not solved. (This, in fact, has proved to be one of the most difficult problems in mathematics, and many famous mathematicians have made contributions to it.)

Book Two treats the motion of bodies in resisting media, such as fluids and air, and Book Three applies the Newtonian dynamics to the motion of the planets, to the theory of tides as caused by the Sun and the Moon, and to various other astronomical questions, such as the trajectories of comets, the motion of the Moon, and the shape of the Earth.

In the *Principia*, the entire program of modern mechanics is formulated, not only in content but also in style. Newton began like a mathematician by first giving definitions of the basic notions, such as mass and momentum, and then formulating three basic laws or axioms from which everything else was to follow:

First law. *Every body remains in its state of rest or uniform motion in the same direction unless it is compelled by impressed forces to change this state.*

Second law. *The change in motion is proportional to the impressed moving force and, secondly, it will occur along the straight line in which that force is impressed.*

Third law. *To an action there is always an equal and opposite reaction, or, the mutual actions of two bodies upon each other are equal and point in opposite directions.*

Although Newton himself deviated from his rigorous approach (in the course of the discussions in his book, he employed undefined notions and applied hitherto unstated axioms), his axiomatic procedure made a strong and lasting impression on all later generations of mathematicians and physicists. Newton had achieved for dynamics what Euclid had achieved for geometry and Archimedes for statics.

Not everything in the *Principia* was new. For instance, the first law is nothing but Galileo's law of inertia, which Newton adopted as one of his axioms. A large part of Book One was taken from earlier writers, but it organized their knowledge systematically. Everything else is Newton's own contribution, and we can only admire how many difficult problems he dared to attack.

[12]

AXIOMATA
SIVE
LEGES MOTUS

Lex. I.

Corpus omne perseverare in statu suo quiescendi vel movendi uniformiter in directum, nisi quatenus a viribus impressis cogitur statum illum mutare.

PRojectilia perseverant in motibus suis nisi quatenus a resistentia aeris retardantur & vi gravitatis impelluntur deorsum. Trochus, cujus partes cohærendo perpetuo retrahunt sese a motibus rectilineis, non cessat rotari nisi quatenus ab aere retardatur. Majora autem Planetarum & Cometarum corpora motus suos & progressivos & circulares in spatiis minus resistentibus factos conservant diutius.

Lex. II.

Mutationem motus proportionalem esse vi motrici impressæ, & fieri secundum lineam rectam qua vis illa imprimitur.

Si vis aliqua motum quemvis generet, dupla duplum, tripla triplum generabit, sive simul & semel, sive gradatim & successive impressa fuerit. Et hic motus quoniam in eandem semper plagam cum vi generatrice determinatur, si corpus antea movebatur, motui ejus vel conspiranti additur, vel contrario subducitur, vel obliquo oblique adjicitur, & cum eo secundum utriusq; determinationem componitur. Lex. III.

Page 12 of the first edition of Newton's *Principia* containing the first two laws of dynamics

The three laws are only the formal framework of dynamics and do not say anything about the nature of the acting forces. In fact, the second law has occasionally been considered tautological. If we want to apply the dynamical laws to concrete cases, we must specify the acting forces. Attraction is one of the basic forces, and Newton stated how this force acts:

> Every particle of matter attracts every other particle with a force proportional to the mass of each, and inversely proportional to the square of the distance between them.

This is known as Newton's *universal law of gravitation*. (Actually, Newton never formulated the law in this general form; instead he gave different versions in different places, which have been combined into the preceding statement.) He named the attractive force of masses *gravitas*, meaning heaviness or weight. Today we speak of *gravity* or *gravitation*.

Hence the gravitation law expresses the difference between mass and weight: a body's mass is independent of its location in space; but if it is placed in a field of gravitation, it experiences weight caused by gravitation. Theoretically, gravitation exists everywhere, but, in practice, the gravitational effect of one body on another is zero if they are far enough apart.

Orbit of Halley's Comet in the solar system.

Newton was able to show that if two bodies attract each other by the inverse-square law, they move on conic sections as trajectories, and the "center of gravity" of the two bodies is situated at a focus of each of these conic sections. If the energy of the bodies is not too large, their orbits will be ellipses, but comets coming from far away could have hyperbolic or parabolic orbits; that is, they would appear once and then go on without ever returning. Other comets with elliptic orbits, such as Halley's comet, return periodically.

Suppose now that one of the two bodies has a large mass and the other has a small mass, like the Sun and the Earth. Then the position of the large body is for all practical purposes identical with the center of gravity, and Kepler's three laws are then a consequence of the law of gravitation.

Newton's philosophy of nature contributed a great deal toward understanding physical phenomena. It showed how these phenomena could be deduced from a few mathematical laws by means of mathematical reasoning. Thus, the old Pythagorean dream of describing the world by mathematics had come closer to fulfillment. Leibniz said of Newton, "Taking mathematics from the beginning of the world to the time when Newton lived, what he had done was much the better half," and Lagrange commented, "Newton was the greatest genius that ever existed, and the most fortunate, for we cannot find more than once a system of the world to establish."

Halley's Comet. June 6 and 7, 1910. Photograph taken by Lick Observatory, California.

Eventually, the agreement between empirical facts and the theoretical predictions of Newton's dynamics became so striking that Newton's theory was accepted as universally true, explaining equally well the motion of the planets and that of a falling stone on Earth. It interpreted the ocean tides as gravitational phenomena caused by the Sun and the Moon, predicted the flattening of the Earth at the poles, and, in one of its most admirable successes, explained the precession of the equinoxes (and of the solstices). The spring and fall equinoxes are the two days in each year when the length of day and night are equal; the solstices are the two days of maximum and minimum length of daylight. These four days, which start the four seasons, slip backward relative to the solar year at a rate of one degree every 72 years. The Greek astronomers had described this phenomenon, but no one had been able to explain it until Newton saw that the Earth's bulge at the Equator is an asymmetry upon which the Sun and Moon, through gravitational attraction, exert a strong influence. This causes the Earth's axis to rotate, tracing out a cone, exactly as a spinning gyroscope may have a moving axis. The Earth's axis completes one revolution approximately every 26,000 years, and this rotation causes the precession of the equinoxes.

In old times, comets were thought to be the harbingers of extraordinary events, either foretelling miracles or, more often, foreboding great disasters, war, famine, fire, flood, and pestilence, the great plagues of humanity. This woodcut taken from J. G. Wagner, *Atmosphera sublunaris*, shows citizens of Augsburg watching a comet. Since Newton's theory was able to explain the motion of comets, these heavenly messengers lost much of their horror, though their appearance in the skies remains exciting to this day.

Head of Brooks's Comet (October 21, 1911). Photographed by Lick Observatory, California.

Until the great success of the *Principia*, Newton's life was quiet and peaceful. He was born on Christmas Day in 1642 (according to our calendar, it would have been January 4, 1643, but at the time the Gregorian calendar had not yet been introduced in England) in the hamlet of Woolsthorpe, Lincolnshire, one year after the death of Galileo. It is said that some attempts to turn him into a farmer failed, and so he entered Trinity College in Cambridge in 1661. For four years he worked quietly, immersing himself in his studies. It has been claimed that during this time Newton acquired his enormous knowledge of geometry by studying the old Greek authors in depth. This seems not to be true, however. D. T. Whiteside, the editor of Newton's *Mathematical Papers*, emphasized that he knows of no material supporting this opinion; Newton had studied neither the *Conics* of Apollonius nor any work of Archimedes; rather he obtained all his knowledge about conic sections from Descartes, through the Latin editions of Descartes's geometry edited by Frans van Schooten (1649 and 1659). Moreover, the variety of results on conic sections in the first book of the *Principia* was a last-minute addendum, inserted

The comet from 1493 depicted in the *Nuremberg Chronicle* (Nürnberger Chronik).

after Newton read the eighth book of de la Hire's *Sectiones Conicae* in 1685.[1] In 1665, Newton received a bachelor's degree at Cambridge. During the two years of the great plague, 1665–1666, Newton remained at home in Woolsthorpe. At this time, according to his own account, he made a large part of his great scientific discoveries. However, this last statement should probably be taken with more than the usual grain of salt.

In 1669, Isaac Barrow (1630–1677), resigned from his position of professor of mathematics at Cambridge in favor of his student Newton, and so Newton became professor, a position he formally held until 1701. After the publication of the *Principia*, Newton became so famous that his energies were increasingly distracted from scientific activities. For two years he sat as the representative of his university in the Parliament at London, and in 1695 he was appointed as Warden of the Mint; four years later he was promoted to the position of Master of the Mint. In 1696, he moved to London and again became a member of parliament. In 1703, he was elected President of the Royal Society, an office he held until his death in 1727.

THE VARIATIONAL PRINCIPLES OF DYNAMICS

We have discussed the development of dynamics at some length in order to show that, initially, variational principles played no role in its creation. But as soon as Newtonian mechanics was accepted on the European continent, the first variational principles were found. In Chapter 1, we looked at the first part of the eighteenth century, when the principle of least action was stated by Euler and Maupertuis. Let us attempt to give a fairly precise version of this principle, as currently accepted, for the simplest case, that of the motion of a single point-mass.

Consider a point-mass m that moves from time t_1 until time t_2 in a *field of conservative forces*, such as gravitation. At each point in a *force field*, a force of a given magnitude and direction acts on each point-mass m moving in the field according to Newton's second law:

A comet's tail always points away from the Sun. (The engraving above is taken from the *Principia*, 1687, and shows different positions of Halley's comet in 1680–1681).

[1] See E. A. Fellmann: "Newton's Principia." *Jahresberichte der Deutschen Mathematiker-Vereinigung* 77 (1975); p. 107–137.

$$\text{force} = \text{mass} \times \text{acceleration}$$

or

$$F = ma.$$

If we visualize a force as an arrow attached to some point, a force field is a system of arrows fixed at all points in space (or of some subdomain of space). We say that such a field of forces F is *conservative* if it possesses a *potential energy U*. This is a rule that attaches a numerical value $U(P)$ to each point P in space, in such a way that the "negative gradient" of these values U at P equals the force F at P. This means the following.

Suppose the function U describes a landscape in a four-dimensional world above the three-dimensional space. (This is, of course, impossible to draw; so the figures here show an energy landscape above a space of only two dimensions.) Let us now consider a point P in space and the point P^* on the energy landscape above P. If P^* is not a stationary point on the mountain range, we can find a direction in the three-dimensional space at P that indicates the *direction of steepest ascent* of the landscape at the point P^* above P; opposite to it, we have the *direction of steepest descent* of the function U. Let us attach an arrow to P, pointing in either one of these two directions, whose size equals the "rate of change" of U in the corresponding direction. In this way, we define two vectors whose feet lie at P. The direction of

(a) Level lines of a mountain above a plane valley, and (b) the corresponding map showing the level lines. The arrows indicate directions of steepest descent at the points where their tail is attached.

A

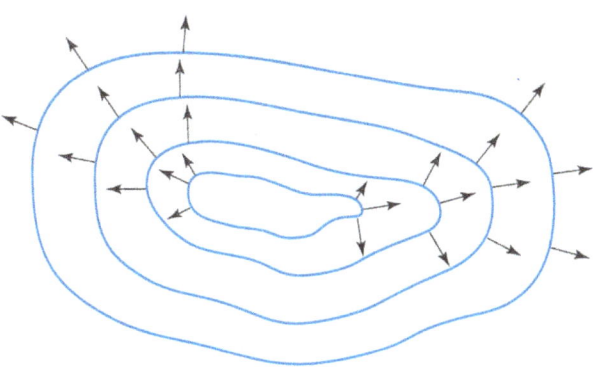

B

steepest ascent defines the "gradient (vector) of U," denoted by grad U, and the vector pointing in the opposite direction defines the "negative gradient," which is the arrow opposite to grad U and therefore is denoted by $-\text{grad } U$.

If P^* is a stationary point, say, the top of a mountain, then the function U has a maximum at P, and there is neither a direction of strongest ascent nor one of strongest descent (this expresses our feeling that, on top of a mountain, the ground is practically horizontal, without ascent or descent). Therefore, we set grad U and $-\text{grad } U$ equal to zero at a stationary point.

Then our assumption that F is a conservative field of forces with the potential energy U is expressed by the equation

$$F = -\text{grad } U,$$

which is to hold at each point of space.

Suppose now that the point-mass m moves with some velocity of absolute value v through space. Since v can change in time, it has to be considered as a function of the time t. Then we can define the *kinetic energy* T of the point-mass at each instant of time by

$$T = \frac{mv^2}{2}.$$

The expression

$$E = T + U$$

is called the *total energy* of the point-mass at each instant.

With some infinitesimal calculus, we can prove from the equations $F = ma$ and $F = -\text{grad } U$ that the total energy E is constant; in other words, a point-mass in a conservative field of forces moves in such a way that its total energy has a numerical value h that is the same at every instant. This is the *law of the conservation of energy*, and it is expressed by the equation

$$E = h.$$

This law explains the term *conservative* applied to a field of forces. We look at two important model cases of conservative fields of forces. The first one is the gravitational field on the surface of the Earth, where one considers the problem of the trajectory of a stone thrown in the air. Here it is assumed that the force of attraction is the same size at each point and that it always points perpendicularly toward the surface of the Earth, which for simplicity is supposed to be a plane. The potential energy U of this field of forces is a linear function of the height h above the ground, and it increases with increasing height. This is described by the rule

$$U(P) = mgh + c.$$

Here m is the mass of the point moving in the gravitational field, c is an arbitrarily chosen gauge constant, and g is a positive constant that, according to measurements, has a value of approximately 9.81 m/sec².

Another model is the gravitational field of a large *mass M*, which rests at some fixed point Q and attracts some point-mass m that moves around Q. The potential energy $U(P)$ of this field at some point P is given by the expression

$$U(P) = -GmM/r + c,$$

where $r = \overline{PQ}$ is the distance of P from the center point Q, G is Newton's gravitation constant, which has, with great precision, been measured by experiments, and c is a number that can be chosen arbitrarily.

The *action A* performed by a point-mass during its motion between times t_1 and t_2 is defined by the integral

$$A = \int_{t_1}^{t_2} 2T\, dt$$

(The symbol $\int_{t_1}^{t_2} 2T\, dt$ on the right-hand side of this formula denotes the area of the shaded domain between the t-axis and the graph of the function $2T(t)$ shown in the figure in the middle of page 291.)

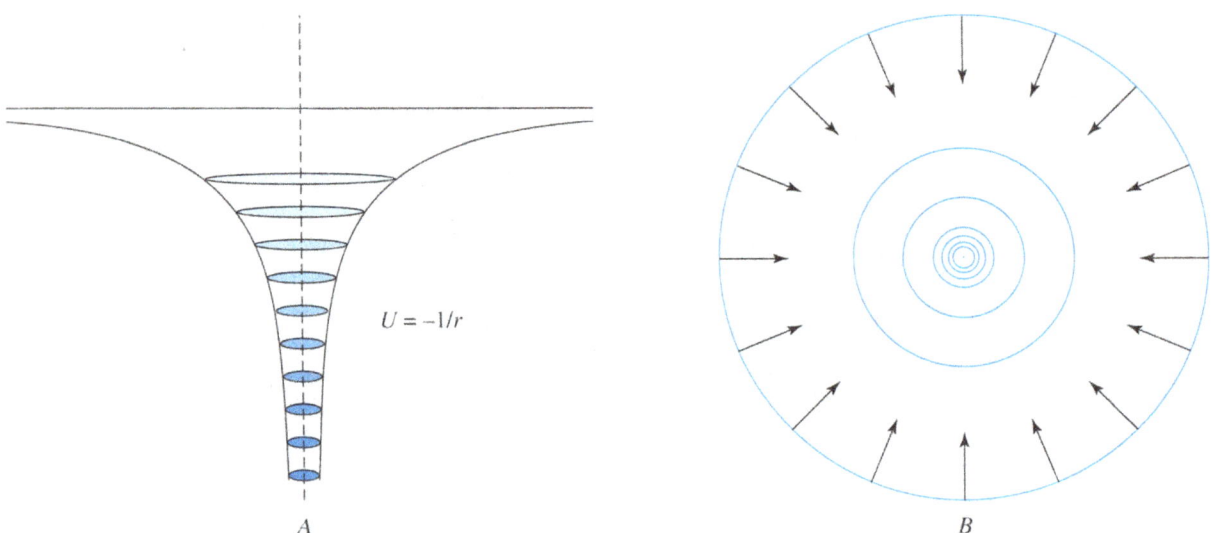

(a) The Newtonian potential of a point mass is proportional to the function $U = -1/r$, whose graph is a surface obtained by rotating a certain hyperbola about a vertical axis which is one of its asymptotes.
(b) The concentric circles indicate the level lines of the Newtonian potential whose picture looks like an (infinitely long) funnel. Since the funnel becomes increasingly steeper as closer one comes to its center, equidistant level lines concentrate more and more toward the center.

Suppose now that the point-mass m moves from a point P_1 to a point P_2 under the influence of a conservative field of forces. What distinguishes the actual motion from all the other motions that, in principle, were possible but actually do not occur?

According to Newton, the actual motion can be determined from the equation $F = ma$ provided that its initial data are known. There is also another way to determine the true motion which is conceptually totally different from the first.

According to Maupertuis, the actual motion of the point-mass m from P_1 to P_2 under the influence of the force field $F = -\operatorname{grad} U$ is distinguished among all other motions by the property that it provides a stationary value for the action A. Actually, here we have replaced Maupertuis's minimum principle,

Seek a motion that minimizes A,

with the following somewhat weaker requirement:

Seek a motion that is stationary for A.

But, as it stands, this principle is pure nonsense, because the acting forces do not appear anywhere in the expression for A. The correct statement would be that we do not need to seek a stationary value for A among all motions from P_1 to P_2, but only among those which at each instant have the same constant total energy E, say, $E = E_0$ as the actual motion. (We need not know the actual motion to compute the *energy constant* E_0; it can, for instance, be obtained from the initial data or from any other complete set of data.)

There is one more complication: in general it is not possible to get from a given point P_1 to another given point P_2 by means of a motion of fixed total energy if we prescribe the initial time t_1 and the final time t_2. Therefore, the demand to minimize the value of action among all possible motions of the point-mass from P_1 to P_2 which have the same total energy E, the same initial time t_1, and the same final time t_2, does not make sense, since there might not be any such motion. The way out of this dilemma is to minimize action among all possible (or virtual) motions from P_1 to P_2 of fixed total energy, for which neither the time of departure nor the time of arrival is fixed.

In this form, the action principle turns out to be correct, but often it is not easy to handle. Therefore, we shall state another version, which was devised by Lagrange. First, we define $L = T - U$, the so-called *Lagrangian* or *action density*; it is the difference between kinetic and potential energy. We can then infer from $E = T + U$ that

$$2T = L + E,$$

and thus we find for motions with a constant total energy of the value E_0 that

$$2T = L + E_0.$$

Now we consider the areas under the graphs of the two functions $2T$ and $L + E_0$ between the two t-values t_1 and t_2. As the two graphs coincide, we infer that

$$\int_{t_1}^{t_2} 2T = \int_{t_1}^{t_2} (L + E_0)dt,$$

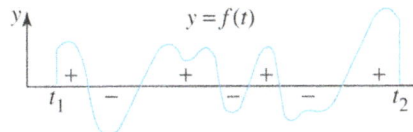

If a function f(t) has both positive and negative values, its integral $\int_{t_1}^{t_2} f(t)\,dt$ between t_1 and t_2 is an "oriented area." The areas above the t-axis are counted as positive, and the ones below as negative. In our example the integral is the sum of the grey areas minus the sum of the blue areas.

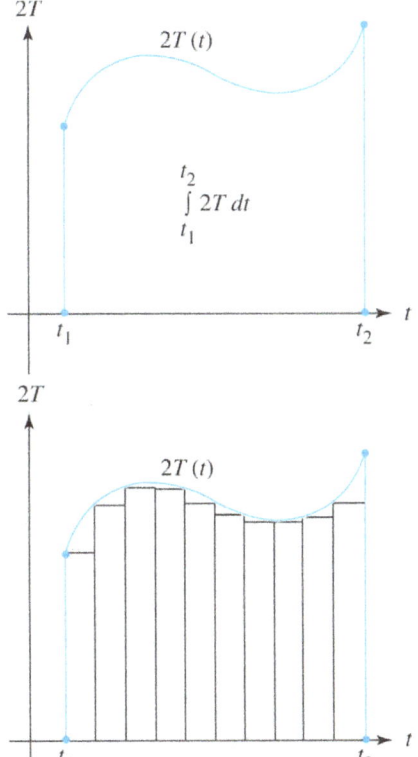

The integral denotes the area of the hatched domain in the figure above. The integral sign \int was introduced by Leibniz and is a stylized S (for sum). This symbol indicates that the area of the hatched domain above can be approximated by the sum of the areas of approximating rectangles (below) if their width tends to zero.

and therefore

$$\int_{t_1}^{t_2} 2T\, dt = \int_{t_1}^{t_2} L\, dt + \int_{t_1}^{t_2} E_0\, dt.$$

The integral on the left-hand side is the action integral A, while the integral $\int_{t_1}^{t_2} E_0\, dt$ is the area of a rectangle of height E_0 above the t-axis which has the interval between t_1 and t_2 on the t-axis as one of its edges. Therefore, this integral has the value $E_0(t_2 - t_1)$, and we obtain

$$\int_{t_1}^{t_2} L\, dt = A + E_0(t_1 - t_2).$$

From this equation we guess correctly that the integral $\int_{t_1}^{t_2} L\, dt$ also can be used to formulate a variational principle for the actual motion from P_1 to P_2. This new variational principle says:

> If the actual motion begins at P_1 at time t_1, and ends at P_2 at time t_2, then it gives the integral $\int_{t_1}^{t_2} L\, dt$ a stationary (and sometimes even a minimal) value, among all motion beginning at P_1 and ending at P_2 at the same time as the actual motion.

The new features of this variational principle are that, contrary to the first one, no *subsidiary condition* $E = E_0$ and no variation of the limits are needed! In other words, the virtual motions are not restricted to those with constant energy E, and we may fix t_1 and t_2. This makes the integral $\int L\, dt$ much easier to handle than $\int 2T\, dt$. In fact, the variational principle

$$\text{Seek a stationary motion for } \int_{t_1}^{t_2} L\, dt$$

is the form of the action principle that can easily be generalized to more difficult situations and to other physical problems.[2]

In Germany and France, this principle was commonly called *Hamilton's principle,* whereas in England the name *principle of stationary*

[2] Heinrich Hertz thought that this variational principle would only hold for a restricted class of mechanical problems; but, suitably interpreted, it is true in general, as was proved by Otto Hölder in 1896.

(*or least*) *action* was used. At present both names are used. Strangely, the principle is not explicitly formulated anywhere in Lagrange's treatise mentioned at the beginning of this chapter, but Lagrange's reformulation of d'Alembert's principle can be identified with the principle "make the action $\int L\, dt$ stationary," justifying our earlier statement that Lagrange based his analytic mechanics on an action principle. This version of the action principle was explicitly formulated by the great Irish mathematician William Rowan Hamilton (1805–1965). He first applied a related principle to *geometrical optics* (i.e., to the optics of *rays*); later he discovered that mechanical problems can be treated in an analogous way. This natural correspondence between problems of point mechanics and geometrical optics was first fully realized by Hamilton, although the origins of this idea can be found in the work of Fermat, Newton, and Johann Bernoulli. At first this observation was nothing more than a formal curiosity and a useful artifice, but in the theories of quantum mechanics and relativity, it led to the understanding of the wave nature of matter, of the particle nature of light, and, eventually, to the equivalence of matter and energy.

Frontispiece of J. L Lagrange's *Mécanique Analytique*, vol. 1, Paris, 1811 (second edition).

TOWARD RELATIVITY

At Newton's time, not everyone was willing to accept the philosophical concepts underlying his theory. We have already mentioned Huygens's skepticism. Johann Bernoulli also remained a Cartesian although he had carefully read the *Principia*. He pointed out numerous errors to Newton, which Newton corrected in the second edition (1713). (The third and final version of the *Principia* was published in 1726, one year before Newton's death.) Bernoulli found it difficult to believe in the concept of a force that acts through the vacuum of space over distances of even hundreds of millions of miles. He viewed this force as a concept revolting to minds unaccustomed to accepting any principle in physics save those which are incontestable and evident. Additionally, Leibniz considered gravitation to be an incorporeal and inexplicable power, philosophically void.

But even Newton was disturbed by questions about how this force acted, and he could give no other explanation than to say, "I have

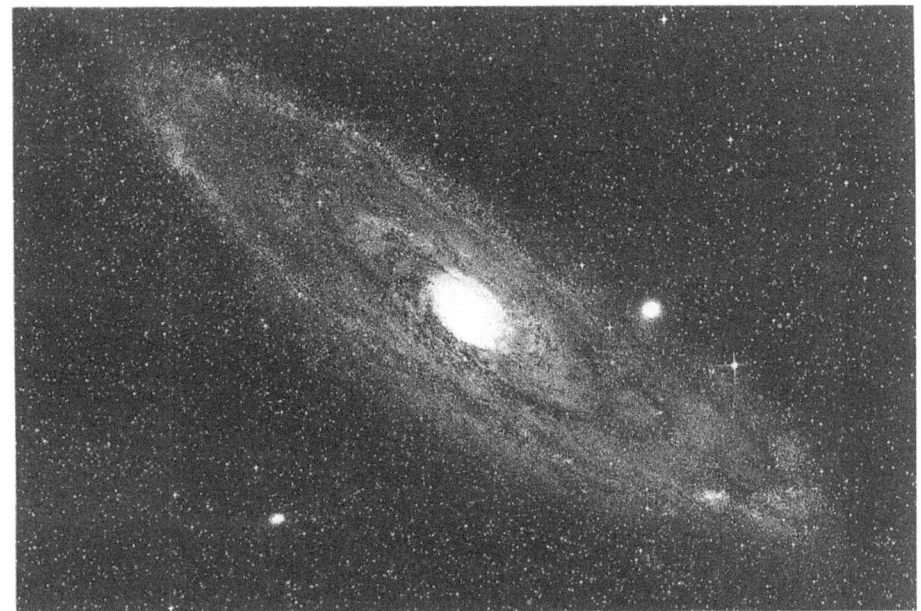

The Andromeda Galaxy. Photograph taken by Lick Observatory (Crossley reflector).

not been able to deduce from phenomena the reason for these properties of gravitation, and I do not invent hypotheses; for anything which cannot be deduced from phenomena should be called an hypothesis."[3] Moreover, in a letter to Richard Bentley, he wrote,

> That gravity should be innate, inherent and essential to matter, so that one body may act upon another at a distance, through which their action may be conveyed from one to another, is to me so great an absurdity that I believe no man, who has in philosophical matters a competent faculty of thinking, can ever fall into it.

Newton coined the term "action at a distance" (which means "force acting at a distance" and should not be confused with the "action"

[3]This is Newton's famous statement in the *Scholium Generale* at the end of the second edition of the *Principia* (p. 484); it cannot be found in the first edition. The original texts reads: "Rationem vero harum Gravitatis proprietatum ex Phaenomenis nondum potui deducere, et Hypotheses non fingo. Quicquid enim ex Phaenomis non deducitur, *Hypothesis* vocanda est; et Hypotheses seu Metaphysicae, seu Physicae, seu Qualitatum occultarum, seu Mechanicae, in *Philosophia Experimenta-li* locum non habent . . . Et satis est quod Gravitas revera existat. . . ."

The Rosette Nebula in red light. Lick Observatory, 20-inch Astrograph.

that appears in the action principle) to describe the mysterious effect of gravitation over large distances. This effect is as difficult to understand today as it was in Newton's time. As we shall see, Einstein was to replace this concept with a more geometric model.

There is an even greater difficulty in Newton's dynamics. It concerns Newton's concepts of an *absolute space* and an *absolute time*. Let us, therefore, have a look at these concepts.

Suppose we ask, "What is space, and how can we find out where we are in space?" Let us see how two of the most famous astronomers determined spatial position.

Galileo measured each position in space with respect to the surface of the Earth, and we still do this in many experiments. This procedure is not correct, but the mistakes made in this way are often negligible.

Kepler employed the astronomical system of coordinates. Its center is fixed at the center of gravity of our solar system, and its coordinate axes are oriented by the fixed stars in the heavens. This is the standard coordinate system of physics that is still in use today.

Yet these two solutions tell us nothing about why one coordinate system is to be preferred to another one. One might answer that it really does not matter theoretically which coordinate system we choose, since any reference system is as good as any other one, and that it really does not matter whether we take this solution or that one, just as it is irrelevant whether we take yard or meter, pound or kilogram, quart or liter, to measure length, weight, and volume. Only practical requirements lead to particular preferences: a carpenter will not measure the lengths of his products in miles or light-years but in feet or meters; and a bus driver would not be happy if the distance from Phoenix to Tucson was indicated in inches or nanocentimeters.

This answer, however, is too simplistic, for the following reason. Assume that Newton's laws hold in some reference system and that, for some reason, we decide to change our coordinate system. For example, suppose we are conducting an experiment in a railway compartment; for practical purposes, we may find it more convenient to choose as a coordinate system any three edges of our compartment that meet in some corner (and measure our position with respect to the corner and the edges) than to employ the geographer's reference system, fixed in Greenwich, England.

Are we then assured that Newton's laws hold in the same form in our compartment system as in the standard system? The answer will be "yes" as long as the train moves straight ahead with uniform speed but will be "no" if the train accelerates; in the latter case, we and our experiment are influenced by additional forces. (To be precise, the train also accelerates in the mathematical sense if it reduces speed; then the acceleration is counted as a negative number. Also, when the train takes a curve with uniform speed, it is accelerating: each change of velocity, be it in absolute value or in direction, means acceleration.)

What can we conclude from this consideration? The laws of dynamics look different in different reference systems, and Newton's laws can therefore hold only in very specific coordinate systems: any two of these systems must move relative to each other with uniform velocity in the same constant direction. What are these "preferred coordinate systems"? Is Kepler's system such a coordinate system, at least with only a very small error?

This shows us that Newton's concept of mechanics requires something like an "absolute space," which we are used to imagining as an infinitely extended three-dimensional box with an absolute coordinate system at rest.[4] Newton's laws hold in this system as well as in all those that move uniformly (no changes in speed or direction) relative to this system. These are called *inertial systems*.

Another concept inherent in Newton's dynamics is the concept of an *absolute time*. This means that once an initial point in time is selected, say, the present, we must then assume that time flows from the infinite past to the infinite future. Moreover, Newton assumed that at any moment we were able to say, at least in principle, where any other object was located in space, whatever our own position might be. In other words, it is always clear whether or not two events occur simultaneously. It was this assumption of Newton that Einstein first shattered in his special theory of relativity in 1905. Because of the experimental data, he was inevitably forced to the conclusion that two events can be simultaneous in one reference system but may not be simultaneous in another system, even if both coordinate systems move uniformly with respect to one another. But it needed an Einstein to draw the correct conclusions from the experimental facts. These facts imply that velocities do not add like vectors (i.e., like "arrows"), as is required in Newtonian dynamics.

[4]The concept of absolute space was rejected by Leibniz, who saw no philosophical need for any concept of space apart from relations to material substance. Leibniz and Samuel Clark, a supporter of Newtonian philosophy, in a famous series of letters, debated this question during the years 1715–1716. Bishop Berkley attacked Newton, whereas later Euler and Kant defended the concept of absolute space.

One of the main insights the special theory of relativity provided is that one cannot separate space and time. Instead, space and time must be viewed as inseparably linked together. According to Einstein, there exist infinitely many inertial systems, all moving with uniform velocity with respect to one another, in which the laws of physics hold. However, if one changes from one system of reference to another, one is not allowed to leave time and geometrical length unchanged, even if two coordinate systems move uniformly with respect to each other. This has the following strange implication.

Suppose we have railway tracks that are in rest in some inertial system and a train that at first does not move. An observer measures the length of the train to be L meters. Then the train starts, accelerates to the speed v, and is measured once more by an observer who looks at the passing train. He determines the length of the train to be $L\sqrt{1 - (v^2/c^2)}$ meters, where c is the speed of light. That is, if the train's velocity v approaches c, its length shrinks to zero for the external observer!

Einstein did not stop at this point. He looked for a formulation of physics where no inertial systems were distinguished but where the fundamental laws were to hold in the same form in each space–time coordinate system. So he tried to find a suitable mathematical setting for his ideas. Assisted by his mathematician friend Marcel Grossmann he realized, with some amazement, that such a setting had already been developed by Gauss, Riemann, and the subsequent

A steel ball on a trampoline.

two generations of geometers from Italy, Germany, and France. Einstein then devised the concept of a four-dimensional space–time world that is not "flat," like the two- and three-dimensional spaces defined by Euclid's axioms, but curved. The curvature of this world is generated by matter and energy. Near a large amount of matter, the amount of curvature of the four-dimensional world is large; far away from matter, it is nearly zero. The exact relation between curvature, on the one hand, and matter and energy, on the other, is the content of Einstein's "field equations." Remarkably, as David Hilbert discovered, these equations follow from a variational principle.

In Einstein's theory, the concept of a force acting through great distances has been replaced by the curvature of a space–time world. In order to elucidate Einstein's scheme, we shall present an oversimplified model that conveys some of his basic ideas. We represent space by a surface that we imagine as an originally flat trampoline (the vacuum state), which is at some point strongly deformed by the weight of a gigantic steel ball (the Sun). A tiny steel ball rolling on the trampoline is our planet Earth.

If we roll the small steel ball across the flat trampoline, it will travel in a straight-line path. However, if we now place the gigantic steel ball in the center, it will cause the trampoline to bend, even "far away" from the large ball. If we then give our little ball a push, it will no longer travel in a straight line, but in a curved path. The big ball affects the trajectory of the little ball by curving the space around it. With just the right push, the little ball might even orbit the big one

Albert Einstein (1879–1955) at his desk in the Patent Office, Bern, 1905.

for a while. This trampoline model explains how a large body could, by curving space, influence a small one over great distance.

Moreover, in Einstein's curved world, light travels along *geodesics*. He predicted that light rays must bend if they pass through areas of large curvature, like the vicinity of the Sun. It is normally impossible to observe light rays close to the bright Sun, but a solar eclipse provides a marvelous opportunity for such measurements. Two British expeditions to New Guinea (under Eddington and Cottingham) and to Sobral in northern Brazil used the solar eclipse of May 29, 1919, to observe whether light rays coming from stars and passing close to the Sun were bent. Both expeditions were able to confirm Einstein's prediction, and Eddington later wrote,

Oh leave the Wise our measures to collate;
One thing, at least, is certain: LIGHT has WEIGHT.
One thing is certain, and the rest debate:
Light-rays, when near the Sun, DO NOT GO STRAIGHT.

We would like to end our account by quoting from a lecture given by Max Planck at the Berlin Academy, June 29, 1922, on "Leibniz Day." After reminding his audience of Leibniz's personality, he drew attention to a very special work by the great scientist and philosopher:

. . . The *Theodicy* . . . culminates with the statement that whatever occurs in our world, in the large as in the small, in nature as in spiritual life, is once and for all regulated by divine reason, and in such a way that our world is the best among all possible worlds.
Would Leibniz reaffirm this statement even today, in sight of the misery of the present time, in view of the bitter failure of many efforts not immediately aimed at material gain, in view of the undeniable fact that the imagined general harmony of people today seems to be further away from its realization than ever? No doubt, we should have to answer this question in the affirmative, even if we did not know that Leibniz never ceased to earnestly occupy himself until his last years despite an adverse

fate and disappointments of all kinds, and we shall hardly err in assuming that it was exactly the *Theodicy* that gave him support and comfort in the most sorrowful days of his life. This once again is a touching example of the old truth that our most profound and most sacred principles are firmly rooted in our innermost soul, independent of experiences in the outer world.

Modern science, in particular under the influence of the development of the notion of causality, has moved far away from Leibniz's teleological point of view. Science has abandoned the assumption of a special, anticipating reason, and it considers each event in the natural and spiritual world, at least in principle, as reducible to prior states. But still we notice a fact, particularly in the most exact science, which, at least in this context, is most surprising. Present-day physics, as far as it is theoretically organized, is completely governed by a system of space–time differential equations which state that each process in nature is totally determined by the events which occur in its immediate temporal and spatial neighborhood. This entire rich system of differential equations, though they differ in detail, since they refer to mechanical, electric, magnetic, and thermal processes, is now completely contained in a single theorem, in the principle of least action. This, in short, states that, of all possible processes, the only ones that actually occur are those that involve minimum expenditure of action. As we can see, only a short step is required to recognize in the preference for the smallest quantity of action the ruling of divine reason, and thus to discover a part of Leibniz's teleological ordering of the universe. The difference in the point of view is, as we see, by no means a difference in essence, but is solely a difference in interpretation, since the first restricts its view to the very small, whereas the other encompasses large distances in space and time. On this occasion everyone has to decide for himself which point of view he thinks is the basic one, and must also ask which approach will eventually be more successful.

In present-day physics the principle of least action plays a relatively minor role. It does not quite fit into the framework of present theories. Of course, admittedly it is a correct statement; yet usually it serves not as the foundation of the theory, but as a true but dispensable appendix, because present theoretical physics is entirely tailored to the principle of infinitesimal local effects, and sees extensions to larger spaces and times as an unnecessary and uneconomical complication of the method of treatment. Hence, physics is inclined to view the principle of least action more as a formal and accidental curiosity than as a pillar of physical knowledge.

Max Planck (1858–1947)

Thus it has been even more surprising that this principle, originally viewed by Leibniz[5] and Maupertuis as a mechanical theorem[6], was found by Hermann von Helmholtz to be valid, without any restriction, throughout the entire physics of his time. Recently, David Hilbert, by employing Hamilton's version of the theorem, has established it in Einstein's general theory of relativity. The more complicated circumstances become, the less likely it is that the dominance of such a simple law could be a mere accident. It is always a sure sign that an approach is incomplete if it cannot explain a commonly accepted, simple, and general relation. Our desire for comprehension will be satisfied only if every law that has been found valid has been understood in all its meaning and relevance, and has been integrated into the entire theoretical structure.

In a talk on a later Leibniz day (June 27, 1935), Planck again deliberated about the main sentence of the *Theodicy*, that among all possible worlds, ours is the best:

Today, when it appears to be more and more difficult to find a reasonable sense in the development of humanity, this sentence may seem grotesque by its audacity. Would Leibniz maintain it in face of the present events, the terrors of the world war, the general mutual distrust and the ever increasing armament of people? We would have to affirm this question, even if we would not know that Leibniz neither lost his philosophy by the horrid ravages of the thirty-years war reaching into the first two years of his life, nor by the subsequent deep humiliation of his country. Namely, it is a matter of fact that the philosophy of the *Theodicy* can neither be proved nor disproved by history. It is not the result of experience, but the supposition and starting point of Leibniz's outlook on the world. According to his philosophical point of view any apparent contradiction, each suffering, every injustice in the world can be traced to the fact that our limited human mind is unable to recognize the whole connection of any single event with the entire plan of the creation. Just as, by distruction of some part, the whole can be advanced, a local evil might help to improve and perfect the general.

[5]Planck considered Leibniz to be the discoverer of the least-action principle.
[6]This statement of Planck is not quite correct, as can be seen from the account in Chapter 1.

Postscript: Planck's oldest son Karl was killed at Verdun in 1916; both daughters died when giving birth to their first child; the younger son Erwin, because he had been involved in the von Stauffenberg assassination attempt of July 20, 1944, was sentenced to death by Hitler's criminal judge Freisler and was executed in January 1945. Shortly thereafter, Planck lost all his personal belongings, and at almost eighty-seven years of age he found himself in the great trek of refugees to the West, just like millions before and after him.

The solar system in cosmos. Engraving by Berot (after F.K. Frisch) in Euler's *Theoria motuum planetarum et cometarum*, Berlin 1744.

References, Comments, and Further Reading

PROLOGUE

Ernst Haeckel (1834–1912), professor of zoology at Jena, Germany, from 1865 to 1909, was one of the first advocates of Darwinism in Germany. He participated in extended scientific excursions throughout the world and systematically investigated and described classes of primitive maritime fauna. His publications, famous for their illustrations, are bibliophilic rarities. A few that are relevant to this book are as follows:

Reports of the Scientific Results of H.M.S. Challenger (London: 1881–1889). Haeckel's contribution to the *Challenger Reports* was reprinted in 1966 (New York: Johnson Reprint Corp.).

Kunstformen der Natur (Leipzig: 1899–1904).

Das Protistenreich: Eine populäre Übersicht über das Formengebiet der niedersten Lebewesen (Leipzig: 1878; Paris: 1880).

Die Radiolarien: Eine Monographie (Berlin: part 1: 1862; parts 2 through 4: 1887–1888.

The following books are expositions, more or less modern, on form and pattern in nature:

D'Arcy W. Thompson, *On Growth and Form* (Cambridge University Press, 1917; abridged edition by J. T. Bonner, 1961).

Peter S. Stevens, *Patterns in Nature* (Little, Brown, 1974).

Wolf Strache, *Forms and Patterns in Nature*, 2d ed. Pantheon, 1973).

Stanislaw Ulam, "Patterns of Growth of Figures: Mathematical Aspects," in Gyorgy Kepes, ed., *Module, Proportion, Symmetry, Rhythm* (Braziller, 1966).

Frei Otto and collaborators: *Natürliche Konstruktionen* (Stuttgart: Deutsche Verlags-Anstalt, 1982).

CHAPTER 1

The life and work of Maupertuis and his place in the history of science have been described by Pierre Brunet in two monographs:

Maupertuis: Étude Biographique (Paris: Blanchard, 1929).

Maupertuis: L'oeuvre et sa place dans la pensée scientifique et philosophique du XVIIIe siècle (Paris: Blanchard, 1929).

The second of these monographs lists Maupertuis's publications and other authors' publications related to Maupertuis's contributions. Moreover, we refer to Brunet's: *Étude historique sur la principe de la moindre action*, Paris, 1938.

A rich source of facts on the history of mathematics and physics in the seventeenth and eighteenth centuries is Leonhard Euler's *Collected Works* (*Opera Omnia*, ser. 1 through 4). Although more than seventy volumes have appeared, the editing is still in progress. Many first-rank mathematicians have added comments and historical remarks to the *Opera Omnia*.

The secondary and tertiary literature on Leibniz and Voltaire is so vast and so easily accessible that we will not list any of it here.

Newton's collected works have been published in this century. See *The Mathematical Papers of Isaac Newton*, D. T. Whiteside, ed. (Cambridge: Cambridge University Press, 1967–1976). The first edition of *Philosophiae Naturalis Principia Mathematica* appeared in 1687, the second edition in 1713, and the third edition in 1725–1726. All later editions appeared after Newton's death. An English translation by F. Cajori, *Mathematical Principles of Natural Philosophy and the System of the World*, was published by the University of California Press (Berkeley: 1934). A definitive edition of the third edition (1726) of Isaac Newton's *Principia*, with variant readings, was published in two volumes by A. Koyré and

I. B. Cohen at Cambridge University Press (1972). An introduction to the *Principia* was published by I. B. Cohen at the same place. E. A. Fellmann's edition of the *Principia* with the transcriptions of annotations by Leibniz (in the copy of the *Principia* kept at the Bibliotheca Bodmeriana) appeared under the title: G. W. Leibniz. *Marginalia in Newtoni Principia Matematica*. Editio prima ab E. A. Fellmann; (in the series L'Histoire des Sciences. Textes et Etudes. Paris, Librairie Philosophique J. Vrin, 1973). The classic biography of Newton is David Brewster's *The Life of Sir Isaac Newton* (London: 1981), which in the enlarged second edition became *Memoirs of the Life, Writings, and Discoveries of Sir Isaac Newton*, two volumes (Edinburgh: 1860). Further biographies of Newton have been written by F. Rosenberger (Leipzig, 1895), A. de Morgan (Chicago and London: 1914), L. E. Moore, *Isaac Newton: A Biography* (1934; available in a Dover reprint), W. Stuckely (1936), and R. W. Westfall: *Never at Rest. A Biography of Isaac Newton* (Cambridge University Press, 1980).

For the history of the calculus of variations, see

Leonhard Euler, *Opera Omnia*, ser. 1, vol. 24.

Herman H. Goldstine, *A History of the Calculus of Variations from the Seventeenth through the Nineteenth Centuries* (Springer-Verlag, 1980), which contains a selected bibliography on the history of the calculus of variations.

The most important source of the history of mathematics since the seventeenth century is the *Encyklopädie der Mathematischen Wissenschaften*, which was edited by the academies of Göttingen, Leipzig, Munich, and Vienna, twenty-four volumes (Teubner, 1898–1934). Two older accounts that are still of interest are I. Todhunter, *A History of the Progress of the Calculus of Variations during the Nineteenth Century* (London: Macmillan, 1861), and R. Woodhouse, *A Treatise on Isoperimetrical Problems, and the Calculus of Variations* (Cambridge, England: Deighton, 1810); both of these have been reprinted by Chelsea Publications, New York.

CHAPTER 2

Despite extensive research on the history of mathematics, our knowledge of the mathematical sciences in antiquity is far from complete. As

an example of a critical attitude toward historical research, we recommend Otto Neugebauer, *The Exact Sciences in Antiquity* (Princeton University Press, 1956).

As an introduction to the mathematics of the ancient Greeks, we suggest the following:

B. L. van der Waerden, *Science Awakening* (P. Noordhoff, 1954).

Morris Kline, *Mathematical Thought from Ancient to Modern Times* (Oxford University Press, 1972).

Morris Kline, *Mathematics in the Modern World: Readings from Scientific American* (W. H. Freeman and Company, 1968).

For scholarly accounts, we refer to the following:

Moritz Cantor, *Vorlesungen über Geschichte der Mathematik,* vol. 1 (Leipzig: Teubner, 1894).

Thomas L. Heath, *A History of Greek Mathematics* (Oxford University Press, 1921).

H. G. Zeuthen, *Geschichte der Mathematik im Altertum und Mittelalter* (Copenhagen: 1896).

Basic translations of Euclid and Archimedes into English are contained in these works:

Thomas L. Heath, *The Thirteen Books of Euclid's Elements* (Cambridge: 1926).

Thomas L. Heath, *The Works of Archimedes with a Supplement: The Method of Archimedes, Recently Discovered by Heiberg* (Cambridge: 1897, 1912).

Both of these translations have been reprinted by Dover Publications, New York. An account of the life of Archimedes, of the history of his manuscripts and editions, and of the historical sources for Archimedes, as well as a new translation of Archimedes' works into English, have been provided by E. J. Dijksterhuis, *Archimedes* (Copenhagen: Ejnar Munksgaard, 1956). Here the reader can find further information on the defense of Syracuse and on the burning mirrors.

A very readable account of the history of astronomy, from the beginning until about 1900, was written by Arthur Berry, *A Short History of Astronomy* (London: John Murray, 1898). This book also contains a well-selected

bibliography of astronomical literature before Einstein's theory of relativity. Another one is R. Wolf, *Handbuch der Astronomie, ihrer Geschichte und Literatur,* two volumes (Zurich: 1890–1893; reprint Hildesheim, 1973).

Concerning the history of the isoperimetric problem, we refer to Helmuth Gericke, "Zur Geschichte des Isoperimetrischen Problems," *Mathematische Semesterberichte* 29 (1982): 160–187.

CHAPTER 3

Heron's principle of light reflection, Schwarz's triangle problem, Steiner's problem, and related questions of the calculus of variations are treated in these monographs:

Hugo Steinhaus, *Mathematical Snapshots* (Oxford University Press, 1950).

Hans Rademacher and Otto Toeplitz, *Von Zahlen und Figuren* (Berlin: Springer, 1933).

Richard Courant and Herbert Robbins, *What Is Mathematics?* (Interscience, 1941; and more recent editions).

D. Hilbert and S. Cohn-Vossen, *Anschauliche Geometrie* (Berlin: Springer, 1932).The first of these is an amusing and easily understood introduction to mathematical questions. The book by Courant and Robbins is the best introduction to mathematics that we know of, but to use it one needs to know some basic mathematics. The last-mentioned book grew out of a celebrated lesson of Hilbert, given at Göttingen University in the winter term 1920–1921. The *Anschauliche Geometrie* is one of the most beautiful books on mathematics but requires a good basic knowledge in mathematics, just like the book of Courant and Robbins.

For the solution of Kakeya's problem, see A. S. Besicovitch, "On Kakeya's Problem and a Similar One," *Mathematische Zeitschrift* 27 (1927): 312–320.

The development of mathematics in the seventeenth and eighteenth centuries is extensively described in

Moritz Cantor, *Vorlesungen über Geschichte der Mathematik,* four volumes (Leipzig: Teubner, 1880–1898).

Publications in English include

F. Cajori, *A History of Mathematics* (New York: 1984; plus various past editions).

D. E. Smith, *A History of Mathematics*, two volumes (Boston: 1923–1925; Dover reprint, 1958).

E. T. Bell, *The Development of Mathematics* (New York: McGraw-Hill, 1945).

W. W. R. Ball, *A Short Account of the History of Mathematics* (New York: Dover, 1960).

C. B. Boyer, *A History of Mathematics* (New York: Wiley, 1968).

E. E. Kramer, *The Nature and Growth of Modern Mathematics* (Princeton University Press, 1981).

In addition, we refer to this work:

Leonhard Euler, 1707–1783: Beiträge zu Leben und Werk, Gedenkband des Kantons Basel-Stadt (Basel: Birkhäuser, 1983).

This memorial volume contains a complete bibliography of the secondary literature on Euler.

Concerning the infinitesimal calculus, sold by Johann Bernoulli to the Marquis de l'Hospital, we refer to *Der Briefwechsel von Johann Bernoulli*, vol. 1 (Basel: Birkhäuser Verlag, 1955): 123–157.

CHAPTER 4

One of the most influential books about the history of mechanics is Ernst Mach, *Die Mechanik in ihrer Entwicklung* (Prague: 1883; and many later editions). Many aspects of Mach's work have been strongly criticized, but for a careful reader the book is still very informative. One should read it together with Clifford Truesdell, *Essays in the History of Mechanics* (Springer-Verlag, 1968).

A survey of the history of the principles of mechanics was given by

István Szabó, *Geschichte der mechanischen Prinzipien* (Basel and Stuttgart: Birkhäuser Verlag, 1977).

René Dugas, *Histoire de la mécanique* (Neuchatel: 1950).

R. Taton, ed., *Histoire générale de la science,* four volumes (Paris: 1957).

The last treatise contains a good bibliography.

The mathematical theory of capillarity was founded by Thomas Young in 1805 (*Essay on the cohesion of fluids,* London Philosophical Transactions of the Royal Society) and Pierre Simon Laplace in 1806 (*Supplément au livre dixième de la mécanique céleste;* see also *Oeuvres* 4, Paris: 1845). Laplace was one of the great scientists of his time. His *Traité de la mécanique céleste* is a cornerstone of celestial mechanics.

For the development of the mathematical theory of capillarity during this century, see Robert Finn, *Equilibrium Capillary Surfaces* (Springer-Verlag, 1986), and for a popular introduction to the field, see the classic by C. V. Boys, *Soap Bubbles, Their Colours and the Forces Which Mold Them* (1911; Dover reprint, 1959).

CHAPTER 5

Giving full credit to the many mathematicians who have contributed to the theory of minimal surfaces is impossible because of lack of space. We will mention only some monographs on minimal surfaces that provide a survey of the field. The theory of minimal surfaces begins with Lagrange's first paper on this subject (1760) and is described in the lectures by Johannes C. C. Nitsche, *Vorlesungen über Minimalflächen* (Berlin, Heidelberg, New York: Springer-Verlag, 1975), and in the treatise by U. Dierkes, S. Hildebrandt, A. Küster, and O. Wohlrab, *Minimal Surfaces I & II* (Berlin, Heidelberg, New York: Springer-Verlag, 1992, two volumes).

Older and much less complete surveys can be found in

T. Radó, "On the Problem of Plateau," in *Ergebnisse der Mathematik und ihrer Grenzgebiete* (Berlin: Springer-Verlag, 1933).

R. Courant, *Dirichlet's Principle, Conformal Mapping, and Minimal Surfaces* (New York: Interscience, 1950).

R. Osserman, *A Survey of Minimal Surfaces* (New York: Van Nostrand, 1969).

The basic ideas on how to use geometric measure theory in the theory of minimal surfaces can be found in F. Almgren, *Plateau's*

Problem: An Invitation to Varifold Geometry (New York: Benjamin, 1966), and in F. Morgan, *Geometric Measure Theory—A Beginner's Guide* (Boston: Academic Press, 1988). The fundamental monograph on geometric measure theory is H. Federer, *Geometric Measure Theory* (New York: Springer-Verlag, 1969). Some recent surveys are as follows:

R. Böhme, "New Results on the Classical Problem of Plateau on the Existence on Many Solutions," *Séminaire Bourbaki* 579 (1981–1982).

A. Tromba, "On the Number of Simply Connected Minimal Surfaces Spanning a Curve," *Memoires of the American Mathematical Society* 12, 194 (1977).

W. Meeks, "Lectures on Plateau's Problem," *Escola de Geometria Differencial* (Univ. Fed. de Ceará, 1978).

M. Struwe, *Plateau's Problem and the Calculus of Variations* (Princeton University Press, Princeton 1988).

F. Jost, *Two-dimensional Geometric Variational Problems* (New York: Wiley, 1991).

A. T. Fomenko and A. A. Tuzhilin, *Elements of the Geometry and Topology of Minimal Surfaces in Three-dimensional Space* (American Mathematical Society Translations, vol. 93, 1991).

L. Simon, *Lectures on Measure Theory* (Proceedings of the Centre for Mathematical Analysis, Australian National University, Canberra, 1984).

We have (nearly) omitted the aspect of optimality that is closely related to the theory of regular configurations, such as mosaics, tilings, densest packings, clouds of spheres, and so forth. This attractive field contains marvelous problems, many of which are unsolved. Fortunately we can refer to an excellent account: L. Fejes Tóth, *Reguläre Figuren* (Leipzig: Teubner, 1965), which is now available in an English translation.

On the optimality of the bee's honeycomb, see L. Fejes Tóth, "What the Bees Know and What They Do Not Know," *Bulletin of the American Mathematical Society* 70 (1064): 467–481.

The biographical material is taken from

J. C. C. Nitsche, "Plateau's Problems and Their Modern Ramifications," *American Mathematical Monthly* 81 (1974): 945–698.

G. van der Mensbrugge, "Joseph Antoine Ferdinand Plateau," *Annuaire Academie Royale Science Bruxelles* 51 (1885): 389–473.

Minimal surfaces, soap bubbles, and other optimal constructions and their applications to architecture are discussed in

Zugbeanspruchte Konstruktionen, Frei Otto, ed., two volumes (Frankfurt, Berlin: Ullstein Fachverlag, 1962 and 1966; an English translation is available).

Ludwig Glaeser, *The Work of Frei Otto* (New York: The Museum of Modern Art, 1972).

Frei Otto and collaborators, *Natürliche Konstruktionen* (Stuttgart: Deutsche Verlags-Anstalt, 1982).

The last book contains a selection of related literature.

The crystallographic point of view is emphasized in Peter Pierce, *Structure in Nature Is a Strategy for Design* (Chatsworth, Calif.: Synestructics, 1978).

CHAPTER 6

The classic on the isoperimetric problem is

W. Blaschke, *Kreis und Kugel* (Berlin: W. de Gruyter, 1916; and later editions).

A survey on results concerning the isoperimetric problem can be found in:

T. Bonnesen and W. Fenchel, *Theorie der konvexen Körper* (Berlin: Springer, 1934; reprinted by Chelsea Publications, New York, 1948).

Phenomena related to the classic isoperimetric problem were studied in the influential monograph by

G. Pólya and G. Szegö, "Isoperimetric Inequalities in Mathematical Physics," *Annals of Mathematical Studies,* no. 27 (Princeton University Press, 1951).

A typical example of a generalized isoperimetric problem is the theorem concerning which plane domain of a given area can support the largest sand pile. This result, as well as the theorem on the cross section of an elastic column with maximal torsional moment, has been proved by

J. Leavitt and P. Unger, "Circle Supports of Largest Sand Pile," *Communications on Pure and Applied Mathematics* 15 (1962): 35–37.

Experiments on capillarity are described in

J. Plateau, *Statique Expérimentale et Théorique des Liquides Soumis aux Seules Forces Moléculaires,* two volumes (Paris: Gauthier-Villars, 1873).

G. Bakker, *Kapillarität und Oberflächenspannung, Handbuch der Physik,* vol. 6 (Akademie Verlagsgesellschaft, 1928).

Some of the results described in the section called "How to Take a Bath in a Spaceship" can be found in these works:

P. Concus and R. Finn, "On Capillary Free Surfaces in the Absence of Gravity," *Acta Mathematica* 132 (1974): 177–198.

P. Concus and R. Finn, "On the Behavior of a Capillary Surface in a Wedge," *Proceedings of the National Academy of Sciences* 64 (1969): 292–299.

J. E. McDonald, "The Shape of Raindrops," *Scientific American* (February 1954: 64–68).

Concerning the section called "Planets, Rotating Drops, and the Nuclei of Atoms," we refer to

J. L. Tassoul, *Theory of Rotating Stars* (Princeton University Press, 1978).

S. Chandrasekhar, "Ellipsoidal Figures of Equilibrium: An Historical Account," *Communications on Pure and Applied Mathematics* 20 (1967): 251–265; and *Ellipsoidal Figures of Equilibrium* (New Haven and London: 1969).

L. Lichtenstein, *Astronomie und Mathematik in ihrer Wechselwirkung* (Leipzig: S. Hirzel, 1923).

L. Lichtenstein, *Gleichgewichtsfiguren rotierender Flüssigkeiten* (Berlin: Springer, 1933).

W. S. Jardetzky, *Theory of Figures of Celestial Bodies* (New York: Interscience, 1959).

W. J. Swiatecki, *The Rotating, Charged, or Gravitating Liquid Drop, and Problems in Nuclear Physics and Astronomy,* International Colloquium on Drops and Bubbles, Pasadena, August 28–30, 1974 (University of California, Lawrence Berkeley Laboratory, LBL 3363).

T. G. Wang, R. Tagg, L. Cammack, and A. Croonquist, *Nonaxisymmetric Shapes of a Rotating Drop in an Immiscible System* (Pasadena, Calif.: Jet Propulsion Laboratory).

On the section called "Cracks and Fissures," see

L. I. Hewes, "A Theory of Surface Cracks in Mud and Lava and Resulting Geometrical Relations," *American Journal of Science* 246 (1948): 138–149.

R. Willden and D. R. Mabey, "Giant Desiccation Fissures on the Black Rock and Smoke Creek Deserts, Nevada," *Science* 133 (1961): 1359–1360.

A. L. Washburn, "Classification of Patterned Ground and Review of Suggested Origins," *Bulletin of the Geological Society of America* 67 (1966): 823–866.

J. T. Neal and W. S. Motts, "Recent Geomorphic Changes in Playas of the Western United States," *The Journal of Geology* 75 (1967): 611–625.

J. T. Neal, A. M. Langer, and P. F. Kerr, "Giant Desiccation Polygons of Great Basin Playas," *Bulletin of the Geological Society of America* 79 (1968): 69–90.

D. E. Derfoot, "Thermal Contraction Cracks in an Arctic Tundra Environment," *Artic* 25 (1972): 142–150.

For the section called "The Geometry of Crystals," see

P. Curie, "Sur la formation des cristaux et sur les constantes capillaires de leurs differentes faces," *Bulletin de la Société de Minéralogistes Français* 8 (1885): 145.

G. Wulff, "Zur Frage der Geschwindigkeit des Wachsthums und der Auflösung der Krystallflächen," *Zeitschrift für Krystallographie* 53 (1914): 171–177.

M. von Laue, "Der Wulffsche Satz für die Gleichgewichtsform von Kristallen," *Zeitschrift für Kristallographie* 105 (1943): 124–133.

A. Dinghas, "Über einen geometrischen Satz von Wulff für die Gleichgewichtsform von Kristallen," *Zeitschrift für Kristallographie* 105 (1944): 304–314.

C. Herring, "Some Theorems on the Free Energies of Crystal Surfaces," *Physical Review* 82 (1951): 87–93.

J. E. Taylor, "Existence and Structure of Solutions to a Class of Nonelliptic Variational Problems," *Symposia Mathematica* 14 (1974): 499–508, Istituto Nazionale di Alti Matematica.

EPILOGUE

The best survey of the development of mechanics, mathematical physics, and special relativity in the nineteenth century is

Felix Klein, *Vorlesungen über die Entwicklung der Mathematik im 19. Jahrhundert,* two volumes (Berlin: Springer, 1926; 1927). This is available in English as *The Development of Mathematics in the Nineteenth Century* (Brookline, Mass.: Math Sci. Pres, 1979).

Further references include

J. L. Lagrange, *Mécanique Analytique,* two volumes, 2d ed. (Paris: 1811; 1815).

C. G. J. Jacobi, *Vorlesungen über Dynamik,* 2d ed. (Berlin: G. Reimer, 1884).

H. Poincaré, "Sur la problème des trois corps et les équations de la dynamique: Mémoire couronné du prix de S. M. le Roi Oscar II," *Acta Mathematica* 13, 1–27 (1889).

On relativity, see

A. Einstein, *The Meaning of Relativity* (Princeton University Press, 1922; many reprints).

H. Weyl, *Space, Time, Matter* (Dover, 1922), translation from the German.

R. W. Clark, *Einstein: The Life and Times* (Avon Books, 1972).

C. Lanczos, *The Einstein Decade, 1905–1915* (Academic Press, 1974).

Concerning Newton, see the references to Chapter 1. An interesting

discussion of the scientific relations between Hooke and Newton has recently been given by M. Nauenberg; "Hooke, Orbital Motion for Central Forces, and Newton's Principia," *Am. T. Phys.* 62 (4) (April 1994): 331–350.

Sources of Quotations

Note: All translations are by the authors except where indicated.

pages vi and vii

Translation by Helen and Hans Lewy from the German original "Korfs Verzauberung," which appeared in Christian Morgenstern, *Palmström* (Berlin: Verlag von Bruno Cassirer, 1922).

page xi

Jonathan Swift, *Gulliver's Travels into Several Remote Nations of the World*, Chapter 2: "A Voyage to Laputa, Balnibarbi, Lluggnagg, Glubbdubdrib and Japan" (first published in 1726).

pages 3 and 48

Aristotle, *Kleinere Schriften zur Physik und Metaphysik*, P. Gohlke and F. Schöningh, eds. (Paderborn, 1957), p. 124.

page 12

Arthur Waley's translation of Hsüan-Tsang's Chinese translation (about A.D. 660) of the Sanskrit *Mahaprajñapāramitā Sūtra*; from *Buddhist Texts through the Ages*, E. Conze, ed. (Oxford: Bruno Cassirer, 1954).

page 16

Albert Einstein, "On the Generalized Theory of Gravitation," *Scientific American*, April 1950.

page 17

P. A. M. Dirac, "The Evolution of the Physicist's Picture of Nature" *Scientific American*, May 1963.

page 19

The Dante quotation is from an essay by R. W. Leonhardt in *Die Zeit*, March 1983.

pages 20, 21, and 35

Pierre-Louis Moreau de Maupertuis, "Le Lois du Mouvement et du Repos, Déduites d'un Principe de Métaphysique," *Mémoires de l'Académie de Berlin*, 1746.

pages 24–26

Voltaire, *Candide or Optimism*, John Butt, trans. (New York: Penguin, 1978).

page 41

Gauss quoting Shakespeare in "W. Sartorius v. Waltershausen," *Gauss zum Gedächtnis* (Leipzig: Hirzel, 1856).

page 63

Johannes Kepler, *Harmonice Mundi* (1619), as translated by M. Kline in *Mathematics and the Physical World* (New York: Thomas Y. Crowell, 1951).

page 83

G. H. Hardy, *A Mathematician's Apology* (Cambridge University Press, 1977), pp. 139–141.

page 85

Dorothy L. Sayers, *Have His Carcase*, Chapter 31: "The Evidence of the Haberdasher's Assistant" (New York: Avon Books, 1968), p. 312.

pages 131 and 226

Pappus, *Collectio*, Book 8. Adapted from E. J. Dijksterhuis, *Archimedes* (Copenhagen: E. Munksgaard, 1956), p. 299.

page 141

Hilaire Belloc, "The Waterbeetle," in *Cautionary Verses*, illustrated by B. T. B. and Nicolas Bentley (New York: Knopf, 1959).

page 148

Richard Courant, *Dirichlet's Principle, Conformal Mapping, and Minimal Surfaces* (Interscience, 1950).

page 159

Gaston Darboux, *Leçons sur la Théorie Générale des Surfaces*, vol. 1 (Paris 1914), p. 490.

pages 226 and 227

D'Arcy W. Thompson, *On Growth and Form* (Cambridge University Press, 1969), pp. 111, 114, and 115.

page 245

"Vous avez confirmé dans les lieux pleins d'ennui ce que Newton connu sans sortir chez lui." The quotation is in S. Chandrasekhar, *Communications on Pure and Applied Mathematics*, 20 (1967): 251–265.

pages 251–253

The editor J. T. Bonner wrote in his Introduction to D'Arcy W. Thompson's *On Growth and Form* (p. vii): "Of its importance there is no doubt, but we must agree with Medawar when he says that its considerable influence has been intangible and indirect." The long quotation on p. 252 is from p. 11 of *On Growth and Form*.

pages 280–283

Isaac Newton, *Principia*, 3d ed. (London, 1726), pp. 13–14. The Newton quotation is from A. Berry, *A Short History of Astronomy* (London: John Murray, 1898), p. 245. Also the Leibniz and Lagrange passages are quoted from A. Berry, pp. 241–244.

page 294

Newton's letter to the theologian Richard Bentley is from *The Correspondence of Issac Newton*, vol. 3 (1688–1694), edited by H. W. Turnbull (Cambridge University Press), pp. 253–256.

page 301

The Eddington quote is from R. W. Clark, *Einstein: The Life and Times* (Avon Books, 1972).

pages 301–303

Max Planck, *Akademie-Ansprachen* (Berlin: Akademie-Verlag, 1948), pp. 41–48 and p. 153.

Sources of Illustrations

pages xiv and 2 (top)

Institut für wissenschaftliche Fotografie, Schloss Weissenstein, Manfred P. Kage.

page 2 (bottom)

Jet Propulsion Laboratory, Pasadena, California.

page 3 (top)

Institut für wissenschaftliche Fotografie, Schloss Weissenstein, Manfred P. Kage.

page 3 (bottom left)

Institut für leichte Flächentragwerke, Stuttgart, Thomas Braun.

page 3 (bottom right)

Eric Pitts.

pages 5 and 6

Institut für leichte Flächentragwerke, Stuttgart, H. C. Heubeck, K. Bach.

page 8

Institut für leichte Flächentragwerke, Stuttgart.

pages 8 (bottom), 9, 10

Institut für wissenschaftliche Fotografie, Schloss Weissenstein, Manfred P. Kage.

pages 11 (top) and 12

Milton van Dyke, *An Album of Fluid Motion*. The Parabolic Press, Stanford 1982. Photographs by Sadatoshi Taneda (p.11) and Magarvay & Latchy (p. 12).

page 13 (top)

Photograph by Michael Meier, Bonn.

page 13 (bottom)

Lick Observatory, California.

page 14

Institut für leichte Flächentragwerke, Stuttgart.

page 15

Institut für wissenschaftliche Fotografie, Schloss Weissenstein, Manfred P. Kage.

page 22

Herzog-Anton-Ulrich Museum, Braunschweig, Museumsfoto, B. P. Keiser.

page 24

Candide, Voltaire, translated by Lowell Blair. Bantam Books.

pages 27–29

Universitätsbibliothek, Basel.

pages 30 (bottom) and 56

Niedersächsische Staats- und Universitätsbibliothek, Göttingen.

page 38 (bottom)

Universitätsbibliothek, Leipzig.

page 39

Forschungs- und Landesbibliothek Gotha, Schloss Friedenstein.

page 40

Museo Archeologico Nazionale di Napoli.

page 43

Mademoiselle M. Houvet, Chartes.

page 47

Lick Observatory.

page 58

Munich Planetarium.

pages 59 (bottom), 60 (top), and 63

Universitätsbibliothek, Leipzig.

pages 60 (bottom) and 61 (bottom)

Universitätsbibliothek, Basel.

page 65

Bayerische Staatsbibliothek, München.

page 68

Hamburger Kunsthalle. Photograph by Elke Walford.

page 70

Biblioteca Apostolica Vaticana, Codex Vergilii Romanus, Nr. 3867.

pge 71

Universitätsbibliothek, Bonn.

page 82

Städtische Galerie, Liebieghaus, Frankfurt am Main.

page 84

Öffentliche Kunstsammlung, Basel. Photograph by Hans Hinz.

page 87 (bottom)

Niedersächsische Staats-und Universitätsbibliothek, Göttingen.

page 89 (bottom)

Konrad Jacobs, Erlangen.

pages 104 (top), and 106

Klaus Vollmer, *Das Steinerproblem auf der Euklidischen Ebene*. Diplomarbeit, Bonn 1991.

page 105

Institut für leichte Flächentragwerke, Stuttgart. Photographs by B. Winkel and J. Kron (Studienarbeit 2/1985 TU Stuttgart).

pages 107 and 108

B. Korte, *Mathematics, Reality, and Aesthetics: A Picture Set on VLSI Chip Design*, Springer-Verlag Heidelberg. Forschungsinstitut fürs Diskrete Mathematik der Universität Bonn.

page 119

Niedersächsische Staats- und Universitätsbibliothek, Göttingen.

Sources of Illustrations

page 124
Bancroft Library, University of California, Berkeley. We thank Michael Nauemberg for printing out this illustration for us.

page 131 (top)
Universitätsbibliothek, Heidelberg.

page 133
Institut für leichte Flächentragwerke, Stuttgart.

page 141 (top), and 142
Institut für leichte Flächentragwerke, Stuttgart, Frei Otto, Klaus Bach.

page 141 (bottom)
Drawing by B.T. B. and Nicolas Bentley (from Hilaire Belloc, "The Waterbeetle," in *Cautionary Verses*, New York: Knopf, 1959).

page 144
The Metropolitan Museum of Art, New York

page 146
Universitätsbibliothek, Bonn.

page 147
Institut für leichte Flächentragwerke, Stuttgart, Klaus Bach.

page 148
©1942 *LIFE* Magazine.

page 157 (top)
Möbius Strip II 1963, by M. C. Escher. Escher Foundation, Haags Gemeetenmuseum, The Hague.

page 159
Konrad Jacobs, Erlangen.

page 160
Department of Mathematics, Columbus, Ohio.

pages 159, 162 and 170
Ortwin Wohlrab, Bonn.

pages 176 and 177
Institut für leichte Flächentragwerke, Stuttgart, Professor Frei Otto.

pages 179, 180 and 186 (top left)
Ortwin Wohlrab, Bonn.

page 183 (top, left), 185 (bottom, left), and 191 (top)
M. J. Bougart, *Zur numerischen Behandlung von Fadenproblemen*. Diplomarbeit, Bonn 1995.

pages 184, 187 (left), and 189
Institut für leichte Flächentragwerke, Stuttgart, Klaus Bach.

page 188
Eric Pitts, Kodak, London.

page 190
Albracht Küster, Bonn.

pages 192 and 193
Institut für leichte Flächentragwerke, Stuttgart, Frei Otto.

pages 194 and 195 (photographs)
Mathematisches Institut, Düsseldorf, Klaus Steffen.

page 196
Eric Pitts, Kodak, London.

page 199
Ortwin Wohlrab, Bonn.

pages 200 and 201
Alan Schoen.

pages 202 (bottom), 203 (right), 204 (top), 205 (bottom), 207, 208, 209, 210 (bottom), 211, 222
Institut für leichte Flächentragwerke, Stuttgart, Frei Otto.

page 212
Museen der Stadt Köln, Schnütgen-Museum. Photo: Rheinisches Bildarchiv.

pages 224 and 229 (left)
Tom Noddy (photographs by Bill Reynolds, Santa Cruz).

pages 225, 228, 229 (right), 230–233
Institut für leichte Flächentragwerke, Stuttgart, H. C. Heubeck, K. Bach, Th. Braun.

pages 234 and 235 (top)
David Hoffman and James T. Hoffman, Amherst.

page 239
Paul Concus and Robert Finn.

page 242
Lick Observatory, California.

pages 242 and 246
Hale Observatories, California.

page 249 (top)
After C. V. Boys.

page 249 (margin)
After T. G. Wang.

page 252
After D'Arcy Thompson.

pages 256 (bottom) and 258
Institut für wissenschaftliche Fotografie, Schloss Weissenstein, Manfred P. Kage.

page 257 (bottom, fig. C)
W. Schuh, Bonn.

page 260
After Derfoot, "Thermal Construction Cracks...," *Artic* 25 (1972): 142–150.

pages 262 and 264 (top)
Steven Smale.

page 263
Egbert Brieskorn.

page 268
National Portrait Gallery, London.

page 278
Emil A. Fellman.

page 282
After JPL, Pasadena.

pages 283 and 285
Lick Observatory, California.

pages 284 and 285
Lick Observatory, California.

page 300
Hebrew University, Jerusalem.

Index

Note: References to illustrations are given in boldface type. A page number followed by the letter "n" refers to a footnote.

Absolute space, 295, 297
Absolute time, 295, 297
Academies
 Berlin, 33, 34, 198n, 300
 French 31, 166
 St. Petersburg, 33, 34
 See also Plato's Academy
Acceleration, 270, 296
Acta Eruditorum, Leipzig, 119, 123
Action, 19–21, 289, 291, 293
 of hiker, 20
 at a distance, 294
Aeneas, 68–71, **68**, **71**
Aeneid, 64, 69, 71
Africa, 64
Agamemnon, 69
Alexander the Great, 72
Alexandria, 72
Alexandrov, A. D., 234
Algae, 2, 4
Algebra, 112
Algebraic formula for minimal surfaces, 174–175
Al-Haytham (Alhazen), **78**
Almagest (al-Magest), 58, **59**, 69n
Altitude triangle, 87–88, 91
Aluminum, **15**
Analytic geometry, 112
Analysis per Quantitatum Series, Fluxiones ac Differentias, 271
Analytical mechanics, 30
Anatase, **263**
Andromeda Galaxy, **294**
Angle(s)
 of 60°, 259, 260, **260**
 of 90°, 109–110, 117, 182–197, 218, 259, 260, **260**
 of 109° 28' 16" 182–197
 of 120°, 3, 5, 93, 95, 97, 100, 102, 182–197
 of incidence, 73, **73**, 74, 76
 of reflection, 73, **73**, 74, 76
 of refraction, 123, 125
Angular velocity, 242, 245, 246
Angular momentum, 241
Annulus-type surface, 177
Antheminus, 81
Apollonius of Perga, 55, **55**, 58, 61, 63, 92, 92n, 285
 Opera (frontispiece), **56**
Aqueducts, 133, 134
Arachnoidiscus USA, 3
Arch, 133
Archimedes, 69n, 73n, 77–83, **79**, 86, 128, 130, 132, 270, 280
 Equilibrium, 130
 Archimedean screw, 79, **81**
 burning mirrors, 77–82, **80**
 theorem of, 79, **79**
Arctic circle, 31, 33
Arctic expedition, **18**, 31, 33
Area, 69, 79, 91
 of a circle, 79
 maximum, 64–71, 212–218
 minimum, 143, 145–160, 206, 226, 231
Aristarchus of Samos, 58
Aristotle, 1, 47, 86
Asia Minor, 48, 69
Astronaut's bathtub, 237–240, **239**
Asymptote, 237, 237n
Athens, 55, 57, 72
Attraction, 274, 275, 279, 282, 284
Augustus, Emperor, 64
Aulosphaera, 254
Axis of rotation, 46, 113, 177, 180, 236
Axopodes, 256

Babylonians, 42, 43, 44, 45
Baker's pretzel, 152, **152**
Ball, 14, 137, 190, 219–222, 241, 246, 250, 299
Ballotini balls, xiv, 3
Baroque period, 17, 21
Barrow, Isaac, 286
Barycenter. *See* Center of gravity
Basel, 21, **28**, 38, 135
Basic unit (of a periodic minimal surface), 198–199
Bath in a spaceship, 237–240
Bee cell, **8**, 226, 227, **227**, **228**
Behaim, Martin, 86
Bending. *See* Isometric deformation
Bentley, Richard, 294
Berkeley, Bishop, 297n
Berlin, 33, 34, 38
Berlin controversy, 33–39
Bern, **300**
Bernoulli, Jakob, 21, 27n, 86, 112, 119, 123, 135, 270
Bernoulli, Johann, 21n, 27, 27n, 33, 38, 39, 87, 112, 117, 119, 120, 123, 125, 135, 135n, 143, 146, 241, 243, 258, 270, 272, 293
Bernoulli's law. *See* Principle of virtual work
Bernoulli's theorem, 117, 118, 120
Besicovitch, 91
Bifurcation, 245–251
 theory of, 248
Binary valley, 250
Bopp, 109
Boracite, **263**
Boundary configuration, 228–229
Boys, C.V., 232
Brachystochrone, 121, 123
 brachystochrone problem, **121**, 121–123
Brahe, Tycho de, 61
Briareus, 80
Bridge, 171–172
British solar eclipse expeditions of 1919, 300
Brook's comet, **285**
Bruno, Giordano, 60

Bubbles, 219–235, 267
 cluster of, 224, **228**
 cubical, **224**
 cylindrical, 223–224, 232
 float of, **14**
 worm of, **228**
Bückner, 98, **98**, 99
Buffon, Georges, 81
Burning mirrors, 72–83, **77**, **78**, 80
Burning surface, 77n

Calcite twinned crystal, **262**
Calculus of variations, 17, 26–30, 89, 138, 175, 215
Callimitra, **7**
Callisto, **2**
Candide, ou l'optimisme (Voltaire), 23–26, **24**
Capillarity, 140, 143, 146
Capillary tubes, 140, **143**
Carleman, Torsten, 219
Cartesians, 31–32
 Cartesian view, **32**, 273, 279
Carthage, 64–71, **65**, **68**, 77
Caryostylos, **255**
Catenary, 135, **135**, 177, **177**
Catenoid, 176–182, **176**, **178**, **179**
Catherine II (tsarina of Russia), 34
Cavalieri, Bonavantura, 92, 93, 112
Catoptrica (Euclid), 73, 73n
Caustic surface, 77n
Cecryphalium, **253**
Celestial bodies, 221, 244
 motions of, 57
 orbits of, 58
Cell, 226, 227, 251–257
Central projection, **53**
Center of projection, 53
Centrifugal forces, 241–251
Chain, **126**, 129, 135, **135**, 177, **177**
Chandrasekhar, S., 248
Characteristic number (of an equilibrium point), 172–175
Chardin, Jean Baptiste Siméon, **144**, 145
Charged drop, 248, 249–251

Chartres cathedral, **43**
Chevalier d'Arcy, 38
Cicero, 79, 82
Cilia, 252
Circle, 1, 46, 52, **53**, **54**, 55, 61, 64, 94, **94**, 163, **163**, 213–218, 236
 center of, 46, 61
Circular motion, 1, 57, 61
Clairaut, Alexis-Claude, 33n, 116, **116**
Clamped plate, 215
Clapeyron, 109
Clark, Samuel, 297n
Classification of topological types, 150–153
Claudianus, 79
Cologne, **67**
Columbus, Christopher, 86
Comet of 1493, **285**
Commentarii of the Imperial Academy of St. Petersburg, 117
Compression, 204
Computer chips, 106–108
 ZORA, **107**, **108**
Concus, Paul, **239**, 240
Cone(s) 43, 53, 79, **79**, 82, 113, **113**, 114, **115**, **118**
 of revolution, 53, **54**
Conformal map, 116
Congruent surfaces, 182
Configuration of minimal area, 226
Conic section(s), 52, 53–55, 61, 75, 282
Conic Sections (Apollonius), **54**, 55
Contact angle, 238, 239, **239**
Convex body, 131
Copernicus, Nicolaus, 58, 59, **59**, **60**, 61, **63**, 244
Cosmogonic hypotheses, 247
Cottingham, 300
Coulomb's law, 242
Courant, Richard, 148, **148**, 186
Courant, R., and Robbins, 169
Cracks, 9, **10**, **13**, 100, 110, 257–261
 mud, **257**
 patterns of, **10**, **13**, 257, 258
 plaque, **258**

secondary, 260
ternary, 260
Critical distance, 178
Croton, 49
Crystalline energy, 266–267, **267**
Crystalline structures, 262
Crystals, **261**, 262–267, **263**, **264**
Cube, 185
Cubical bubble, **224**
Curvature, 115, 162–167
 of a curve, 162–163
 Gaussian, 115, 165–166
 mean, 165–166, 225, 231, 234, **234**, 235, 236
 of a surface, 165–166
 total, 169
Curve(s), 112
 closed, 64, 147, 168, 170–172
 of constant curvature, 206
 convex, 217, **217**
 interlocking, 182, **182**
 knotted, 160
 length of, 64
 nonconvex, 217
 planar, 74, 162
 probability of choosing a bad one, 172
 of quickest descent, 121, 123
 random choice of, 172
 with self-intersection, 158, **234**
 without self-intersection, 158
 of shortest length, 85–125
 simple, 154, 158
 smooth, 172, 237
Cusp, 77n
Cycloid, 52, 53, **53**, **121**, 121–125
Cylinder, 43, 79, **79**, 82, 111, **111**, 113, **114**, 116, 118, 231, 235, 236
Cylindrical soap bubble, 223–224, 231
Cyrtocalpis, **253**

d'Alambert, Jean le Rond, 272, 293
da Cremona, Jacapo, 130
Dante Alighieri, 69

Darboux, Gaston, 159
Darwin, George, 248
De Boissieu, Jacques, 221
Delaunay, C., 236
De linea brevissima (Euler), 116, 117n
De Prima, Charles, 148
De revolutionibus orbium coelestium (Copernicus), 59
Deformation without distortion of lengths, 114, 115
Deformed bowl, 161, **161**, 173–174
della Bella, Stefano, **61**
de Roberval, 274
Descartes, René, 31, **31**, 112, 125, 273, 274
Devil's Post Pile, 13
Dialogue Concerning the Two Chief World Systems (Galileo), 60
Diatoms, **4**
Diatribe of Doctor Akakia (Voltaire), 38, **38**
Dido, 64–71, **65**, **68**, **70**, **71**, 213
Dido and Aeneas (Purcell), 69
Dido's problem. *See* Isoperimetric problem
Die Akustik, 216
Differential calculus. *See* Infinitesimal calculus
Differential equation(s), 30, 117, 125
Dinghas, Alexander, 267
Dirac, P.A.M., 17
Direct method in the calculus of variations, 89
Dirichlet, Peter G. Lejeune, 218, 248
Dirichlet's principle, 137
Disc, 235
Disc-type minimal surface, 154–155, 159, 168–175
Disquisitiones generales circa superficies curvas (Gauss), 115
Divine Comedy (Dante), 69
Doctor Faustus, 51
Dodecahedron within bubble cluster, **224**
Double helix, 180 **180**
Doughnut, 152, **152**

Douglas, Jesse, 159–160
Dragonfly, 6
Drop(s), 9, **9**, 141–143, 232
 pendant, **141**, 240
 sessile (or sitting), **141**, 240
Drop model of nucleus, 249–251, **250**
Dynamics, 269–304
 definition of, 271

Earth, 57, 58, 61, 284, 289, 295
 axis, 284
 flattening of, 31, 245, 284
 shape of, 32, 86, 221, 241–248
Earthquake of Lisbon, 25
Eddington, Sir Arthur Stanley, 300
Edge of a polyhedron, 174, 253
Egyptians, 42, 44, 45
Einstein, Albert, 272, 273, 297, 298, 299, 300, **300**
 Einstein's field equations, 272, 299
Elastic forces, 141–143
Elastic skin, 140
Electric forces, 140, 243, 244
Elements (Euclid), 72
Ellipse, 53, **53**, **54**, 55, 61, 63, 76, **76**, 94, 236, 245n
Ellipsoid, 111, 113, 245, 245n
Elliptical orbit, 61
Empedocles of Agrigentum, 72
Energy, 20, 244, 287, 288
 crystalline, 264–267, **267**
 electrostatic, 244
 gravitational, 143, 244
 kinetic, 139
 minimal, 139, 142, 219
 potential, 137–142, 146, 160, 172–175, 264–267
 rotational, 244, 248
 surface, 140–143, 244, 248, 264–267
Energy barrier, 249–250
Energy landscape, 250, **250**
Enneper surface, 162
Epicycle theory, 27
Epicycloid, 53, **53**

Epitrochoid, 53
Equilibrium (Archimedes), 130
Equilibrium, 126–143, 238, 246, 248 250
 gyrostatic, 244, 246
 stable, 134, **134**, 135, 137, 142, 160, 204, 219, 244
 unstable, **134**, 135, 137, 160
Equilibrium figure of a liquid, 219, 242–251
Equilibrium position (point), 143, 172–175, 219
Equilibrium state, 138, 146, 160, 238 250
Equinoxes (precession of), 284
Eratosthenes, 86
Escher, M.C., 156, 157
Euclid, **55**, 69, 72, **72**, 73, 130, 271, 280, 299
Eudoxus of Cnidus, 57
Euler, Leonhard, 27, 29, 30, 33, 34, 35, 39, 63, **84**, 112, 116, 117n 117, 119, 120, 127, 146, 174, 178, **271**, 272, 286, 297n
 Euler-Lagrange equations, 30
 Euler's formula, 174
Europa, **2**
Evolute, 123, **123**
Existence of a solution, 88, 119, 148–158, 218
Eye loop, 203, 207, **207**, **208**, 211, **211**

Faber, 215
Fabricius, 244
Face of a polyhedron, 174
Falling bodies, law of, 61, 123
Fasbender, 100, **100**
Fejes Tóth, L., 227, 228
 Fejes Tóth's cell, **227**
Fellmann, Emil A., **278**, 286n
Fermat, Pierre de, 37, 92, 112, 123, **123**, 293
 Fermat points, 92–100
 Fermat's principle, 37, 123, **123**
 Fermat problem, 92

Fields medal, 160
Finiteness theorem, 172, 175
Finn, Robert, **239**, 240
Fission, 249–251, **247**, **250**
Flagella, **2**
Flagellates, **2**, 3
Fleeting ghost, 161, 168
Float of soap bubbles, **14**
Fluorite, **263**
Fluxions, 271
Foam, 229
Focus
 of a conic section, 55, **56**, 61, 76, **76**, 77, **77**, 236
 roulade of, 236, **236**
Fontenelle, 228
Force, 221, 242, 248, 252, 270, 271, 273, 274, 275, 276, 280, 282, 286, 287, 288, 289, 293, 294, 296, 299
 conservative, 286, 288, 289
Fortuna, 219, **219**, 220
Fractures, 257–260
 patterns of, 9, **13**, 257–260
Framework, 146, 182, 194–196
 octahedral, 196, **196**
 quadrilateral, 197, **198**
 tetrahedral, 192, 195
Frederic II, King of Prussia, 33, 34, 146
Free boundary, 183–191, **183**, **185–189**
 free-boundary problem, 183–191
Free surfaces, 238
French Academy of Sciences. *See* Academies, French
Friction, 248, 249
Frustum of a pyramid, 45, **45**

Galen, 80
Galilei, Galileo, **2**, **60**, 60–61, **61**, 63, 121, 244, 270, 273, 280, 285, 295
 Discorsi e Dimostrazioni Matematiche, **138**
 law of a falling body, 60–61, 123
 law of inertia, 280
Gamov, George, 249

Ganymede, **2**
Garey, 104
Gauss, Carl Friedrich, 41, 86, 109, 115, **115**, **130**, 143, 298
Gaussian curvature, 115, 165–166
Gechauff, Thomas, 130
General theory of relativity, 293–300
Generalized Steiner problem, 100–109, 142
Genus of a surface, 150, 152, **153**, 153, 154–156
Geocentric system, 58
Geodesic, 118–120
Geodesic network, 193, **194–195**
Gergonne, J.D., 109, 185
 Gergonne's surface, 185, **185**
Giant's Causeway, **261**
Goldbach, Christian, 149
 Goldbach conjecture, 149
Graph over a sphere, 267
Gravitation, 31, 63, 121, 219, 237, 282, 284, 289, 293, 294, 295
 center, of, 130–131, 135, 219, 282, 283
 gravity, **239**, 240, 241
 law of, 221, 241, 282, 283
Great circle, 46, 118, 193, **194–195**
Greece
 geography, 86
 schools, 52
 science, 41–83
Greenwich, England, 296
Grossmann, Marcel, 298
Grossularite, **263**
Gutbrod, Rudolf, 203
Gyroscope, 242

Haeckel, Ernst, **7**, **8**, 253, **254**, **255**, **256**
Halley, Edmond, 276, **276**, 277, **277**, 278
Halley's comet, 63, 282, **282**, **283**, 286
Halo, **9**
Hamilton, William R., 63, 125, 292, 293, 302
Handle, 151, **151**

attachment of, 151
Handle-body surface, 151–155
Hardy, G.H., 83
Harmonious chords, 48
Harmony, 41, 48
Harmony of the World (Kepler), 63
Harriot, Thomas, 244
Hattendorff, 198
Heinen, F., 99
Hekademeia (Plato's Academy), 57
Helicoid, 176–182, **179**, **180**, **181**
Heliocentric system, 58, 244
Helix, 114, **114**, **118**, 119, 211, **211**
Helmholtz, Hermann von, 302
Hemisphere, 29, 174
Henneberg, 190
 Henneberg surface, 190, **190**
Henzi, Samuel, 38
Heron, 74
 law of reflection, 74–75
 principle of, 74, 87
Hessonite, **263**
Hexagon, 225–228, **225**, **226**, **227**, 256, 258, **259**
Hexagonal packing, **6**, 225–228
Hexahedrons in a bubble cluster, 224
Hilbert, David, 119, **119**, 299, 302
Hipparchus, 58
Hippasus of Metapontum, 50, 51
Hippocrates, 80
Holes in a surface, 151–153
Honeycomb, **6**, 226–228, **226**, **227**, **228**, 256
Hooke, Robert, 274, 275, 276, 277
Hopf, Heinz, 234
Horologium Oscillatorium (Huygens), 122, **122**, 274
Horse-watering problem, 74–75, **74**
Hourglass figures, 250, **250**
H-surface of revolution, 235–237, 252, **252**
Huygens, Christian, **121**, 121–122, **122**, 270, 273, 274, 293
Hydraulics, 130
Hyperbola, 54, **54**, 55, 236, **236**, 237n, 240

Hyperbolic cosine, 135
Hyperboloid, **111**
Hypocycloid, 53, **53**, **91**, 91
Hypotrochoid, 53

Igloo, 29
Imperial Russian Academy of Sciences. *See* Academies, St. Petersburg
Inclined plane, 128–130, **128**
 law of, 129
Index, papal, 26, 59
Infinitesimal calculus, 26, 27, 119–120, 138, 163, 270, 271, 272
Inertia, law of, 280
Inertial systems, 297, 298
Inquisition, 60
Inscribed triangle, 87, **87**, 87n
Integer, 47, 48
Integral calculus. *See* Infinitesimal calculus
Interlocking curves, 182, **182**
Internal motion, 240, 248, 249, **249**
Involute, 123
 of a cycloid, 123
Io, **2**
Irrational numbers, 50–51
Isochrone, 122
Isometric deformation, 114–115, 179, **179**
 of a catenoid into the helicoid, 179, **179**
 See also Mapping without distortion of length
Isoperimetric inequality
 of the ball (sphere), 221
 of the disc (circle), 214
Isoperimetric problem, 64–71, **66**, **67**, 213–224, 264
 for crystals, 264–267
 for honeycombs, 226–228
 for reduced surface area, 223
Isoperimetric property
 of the circle, 69n, 213–218, 224, **224**
 of the sphere, 221–222

Isosceles triangle, **49**
Istoria e dimostrazioni (Galileo), 244

Jacobi, Carl Gustav Jacob, 63, 246
 Jacobi ellipsoids, 246–247
Jarbas, King of Numidia, 64
Jerusalem, 86
Jet Propulsion Laboratory, Pasadena, 249
Johnson, 104
Jupiter, **2**, 46
Justinian, Emperor, 57
Justinus, 65

Kakeya's problem, 91
Kant, Immanuel, 247, 297n
Kepler, Johannes, 52, **57**, **61**, 61–63, **62**, **63**, 270, 273, 283, 296, 297
 first law, 61, 283
 second law, 63, **63**, 283
 third law, 63, 275, 283
Kinematics, 270
Kinetic energy, 20, 139
Klein, Felix, 158
 Klein's bottle, 158, **158**
Klingenstierna, Samuel, 120
König, Johann Samuel, 33n, 38, **38**, 226
Krahn, 215
Kristan von Hamle, 131

La Condamine, C., 32
Lagrange, Joseph Louis, 30, **30**, 39, 63, 146, 166, 246, 272, 283, 291, 293, **293**
Lambert, 50
Lamé, 109
Laminae of minimal potential energy, 146
Laplace, Pierre Simon, Marquis de, 143, 247, 278
 Laplace equation (for the pressure at an equilibrium surface), 167
Lappland, 31, **31**
Laws
 of collision, 273

 of conservation, 273
 of conservation of energy, 288
 of falling bodies, 60–61, 123
 of gravitation, 221, 241, 276, 279, 282, 283
 of the inclined plane, 129
 of inertia, 280
 of least action. *See* Principle of least action
 of motion, 19, 280, **281**
 of the osculating plane. *See* Bernoulli's theorem
 of reflection, 73n, 73–76, 94, **94**, 123n
 of refraction, 123, 125, **125**
 of rest, 19
Le bulles de savon, 221
Leibniz, Gottfried Wilhelm von, 21–23, **22**, 26, 33, 38, 39, **39**, 112, 123, 135, 270, 271, 278n, 279, 283, 293, 297n, 300, 301, 302
Length
 of a curve, 64
 minimum, 85–125, 197
Leonardo da Vinci, 135, **136**, 140
Lever, 130
Lewy, Hans, vi–vii, 186
Lewy, Helen, vi–vii
Lichtenstein, Leon, 248
Liebmann, Heinrich, 267
Light
 path of, 35, 36, **91**
 propagation of, 35–36
 ray(s), 72–73
 speed of, 72
 triangle. *See* altitude triangle
Lindemann, 50
Liquid bodies, 24–249
Liquid edge, 191–197, 225, 253
Liquid skin, 140, 141
L'Hospital, Marquis de, 120, 123
 L'Hospital's rule, 120n
Loch Ness monster, 154, **155**
Loop, 188, **188**, 203, 207, **207**, **208**
Louvre, 146

Lyapunov, Alexander S., 248, **248**

MacLaurin, Colin, 245, 246, 247
 MacLaurin ellipsoids (spheroids), 245–247
Magic Flute, 134
Magnification procedure, 192–195
Magnetic forces, 140
Manesse manuscript, 131
Manhattan–Steiner problem, 106
Mapping without distortion of length, 114–115
Marcellus, 80, 82
Marchese del Monte, Guidobaldo, 135, **135**
Marquise du Chatelet, 33, 33n, **33**
Mars, 46, **58**, 62
Mass, 20, 280
Massica, **239**
Mathematical model(s), 16, 17, 19–21, 46, 48, 57, 59–60, 73, 125, 130, 137–143, 147, 231–234, 299, **299**
Maupertuis, Pierre-Louis Moreau de, **18**, 19–21, 26, 30–39, **31**, **38**, 116, 286, 290, 302
 Maupertuis's principle. *See* Principle of least action
Maximum, 29, 39, 100
Mean curvature, 165–167, 225, 231–232, 233–235
Mechanics, Newtonian, 30–32, 269–297
Mécanique analytique (Lagrange), 30, 272, **293**
Mécanique céleste (Laplace), 143
Mediterranean Sea, 44, 64
Membrane, 215
 tension, 253
Mémoire de la courbure des surfaces (Meusnier), 166n
Menaechmus, 55
Mercator, Gerardus, 86, **86**
 Mercator projection, 116
Mercury, 46, **58**, 62, 240
Meridian, 113, **113**, 114

Mesopotamian mathematics, 44
Metaphysics, 16, 19
Metaphysics (Aristotle), 47
Metapontum, 52
Methodus inveniendi (Euler), 35, **36**
Meusnier, J., 166n, 180
Microtubuli, 256
Miletus, 42
Minimal connection, 102, 106
Minimal perimeter enclosing prescribed area, 214
Minimal system of paths, 100
Minimal surface area enclosing prescribed volume, 223
Minimal principal frequency, 215
Minimal surface(s), 143, 145–211, 231, 232
 algebraic formula for, 172–175
 bounded by threads, 205–211
 of disc-type, 154–155, 158–159, 168–175
 with a free boundary, 182–197, 208
 number of, within a contour, 168–172
 with a partially free boundary, 182–183
 periodic, 197–201
 really different, **182**, 182, **185**, 185, 196
 of revolution, 177–178
 stable, 145–160, 161, 167, 170, 172, 176, 204
 unstable, 160–168, 175
Minimal surface equation, 167
Minimum, 28, 29, 39, 85–125, 137–143
 absolute, 250, **250**
 local, 160, 250, **250**
Minimum area, 143, 145–160, 226, 227, 231
Minimum length, 85–125, 197
Minimum of potential energy, 127, 137–143, 146, 219, 244, 250, 264
Minimum principle, 37, 137, 146, 219, 226, 264–265, 290

Minimum problem, 226–227, 264
Minimum property, 46, 214, 215, 219, 226, 250, 264
Minkowski, Hermann, 272, **272**, 273, **273**
Mirror(s), 37
 concave, 37
 convex, 37
 curved, 73, **73**
 flat, 73–74
 hexagonal, 80–81
 parabolic, 77, 81
 plane, 73–74
 spherical, 37
Mirror image, 90, 199, 217, **217**
Möbius, August Ferdinand, 156
 Möbius strip, 156–158, **157**, **158**
Molecular forces, 142
Momentum, 280
Monster curve, 155, 171–172
Montreal, German Pavilion, 202, 208
Moon(s), 46, 57, 62, 240, 242, 284
 of Jupiter, **2**
Morse, Marston, 174
 Morse's equation, 173, 174
Motion, 269–303
 apparent, **58**
 circular, 1
 of heavenly bodies, 47–57, 269–303
 law of, 19
 of planets, 34, 57, 61, 62, 269–303
 spherical, 62
Mountain pass, 29, **30**, 62, 250
Mountain-pass lemma, 172–175, 250
Mountain range, 29, 30, 137, **137**, 250
Multiply connected, 241
Munich, 203–205, 208
 Olympic Stadium, 202, 203, 204
 Olympic Athletic Arena, 203
 Olympic Swimming Arena, 204
Museum of Alexandria, 72, 79
Music of the spheres, 46, 48

NASA drop tower, 239, 240

Nauenberg, Michael, 272, 276n
Nebulae, 244
Neil's parabola, 190, **190**
Neovius, 186
 Neovius's surface, 201
New Guinea, 300
Newton, Sir Isaac, 26, 27, 30, 31, 63, 119, 123, **268**, 269–297
 constant of gravitation, 246
 law of gravitation, 221, 241
 laws of motion, 280
 Propositiones de Motu, 276, 277
 view of the Earth, 31, **32**
Newtonian physics, 30–32, 39, 245
Nodal point, 92, 102, **106**
Noddy, Tom, **224**, **229**
Nodoid, 235, **236**, 237
Nonexistence of a solution, 176, 210
 to Kakeya's problem, 91
Nonwetting liquid, **143**, 240
Normal direction, **265**, 266
Normal plane, 164, **164**
Normal section, 164, **164**
Notch effect, 259, 260
Nouvelle mécanique (Varignon), 135n
Nuclei of atoms, 249–251

Obstacle problem, 186–187, 208–210
Oil drop, 14, 248
On Floating Bodies (Archimedes), 130
On Growth and Form (D'Arcy Thompson), 226, 251–257
On the Equilibrium of Planes. See Equilibrium (Archimedes)
Opera Omina (of Euler), 34
Optics (Euclid), 72–73
Optimal shape, 28, 204
Optimum principle, 42, 63, 72, 226
Orbit(s), 270, 274, 282
 circular, 61
 elliptic, 61, 63, 276, 282
 hyperbolic, 282
 parabolic, 61, 63, 282
Oscillation of nucleus, **250**
Osculating plane, 117, **117**, 118
Otto, Frei, 202–205, 207–210

Palladium, 15
Pangloss, 23–25
Pantheon, 132, **133**
Pappus, 69n, 131, 226
Papyrus Rhind, 42
Parabola, 54, **54**, 55, 61, 76, 77, **77**, 230
Paris, **66**
Paraboloid, 76–77
Partially free boundary, **183**, **184**, **185**, **187**
Pascal, 33
Peak, 29, **30**
Pear-shaped body, 247, **247**
Pendulum, 121–123, **122**, **123**, 274, 279
 perfect, 122–123, **123**
Pendulum clock, 121–123, **122**
Pentagon, 258, **259**
Pentacle (star), 50, **50**
Pentagram, **50**
Pericles, 55
Periodic minimal surface, 197–201
Perimeter, 69, 87, 87n
Permafrost, 258
Perpendicular, 109, 110, 117, 137, 183, **183**, 199, 237, 265, **265**, 266
Petreius, Johannes, **60**
Peurbech, Georg von, **55**
Philosophiae Naturalis Principia Mathematica. See Principia (Newton)
Pi, 42, 50, 79
Piecewise smooth body, 264, **264**
Pit, 30, 250
Planck, Max, 300, 301, **301**, 302, 302n, 303
Plane, 177, 237
 of incidence, 73
 of reflection, 73
Planetarium, 79
Planetary motion, 34, 57, 58–61, 270–285
Planets, 1, 57, **58**, 221, 241–248, 267, 270, 273, 275
Plaque, **258**
Plateau, Joseph Antoine Ferdinand, 146, **146**, 191, 232, 248, 249, 253
 experiments with rotating drops, 248–249, **248**, **249**
 problem of, 146–160, 168, 169, 172
 problem for disc-type surfaces, 158–159
 Plateau's rules (Rules 1 and 2 for soap films with free boundaries), 183, 192, 225, 253
Plato, 39, **40**, 52, 57
Plato's Academy, 57
Plato's school, 55, 57
Platonic bodies, 51
Plutarch, 80, 83
Poincaré, Henri, 247, 248
 Poincaré bifurcation, 247, 248
Poles, 31, 245
Polygon, 159, **159**, 257, 258
Polyhedron, 14, 174
Pope, Alexander, 279
Possible states (virtual states), 128, 143
Potential energy, 127, 137–143, 146, 160, 172–175, 264–267, 287, 288, 289
 minimum of, 137–143, 146
 Newtonian potential, 287, 288, 287, **290**
Prague, 61
Preestablished harmony, 22, 23
Pressure, 225, 232
 difference in, **4**, 167, 232
Pretzel surface, 152, **152**, 241
Prime numbers, 149
Principal axes of an ellipsoid, 245
Principal curvatures, 166
Principal radii of an ellipsoid, 245
Principia (Newton), 31, 33n, 245, 270, 271, 272, 273, 277, **277**, 278, 279, 280, 281, **281**, 285, 286, **286**, **294**
 frontispiece, 277
Principia generalia theoriae fluidorum in statu equilibrii (Gauss), 143
Principle of economy of means, 16, 17

Principle of greatest economy. *See* Principle of least action
Principle of least action, 19–21, 26, 30, 33–39, 273, 286, 291, 292, 301, 302, 302n
Principle of steepest ascent or descent, 287, **287**, 288
Principle of virtual work, 135, 136, 146, 219, 243, 258, 272
Prism, 128, **129**
Problem(s)
 of Queen Dido. *See* Isoperimetric problem
 of quickest descent, 120–125, **121**
 of shortest connection, 85–120
Proclus, 69, 213
Projection, 168, **168**
Prussian Academy of Sciences. *See* Academies, Berlin
Ptolemaic theory, **55**, 58, **58**, 63
Ptolemy (of Alexandria), 57, 58, **61**, 69n
Ptolemy Soter, King of Alexandria, 72
Pulley, 130, **131**
Punic wars, 77
Purcell, Henry, 69
Pygmalion, King, 64
Pyramid, 14, 43
 frustum of, 45, **45**
Pyrite, **263**
Pythagoras, 42, **43**, 44, 46–52
Pythagorean model of the heavens, 46, 47, 48, 283
Pythagorean physics, 47, 48, 215, 283
Pythagorean school, 49–52
Pythagorean theorem, 43, 46, 49, **49**, 97

Quadrilateral, 197, **198**, 200
Queen Dido's problem. *See* Isoperimetric problem
Quickest connection, 85–125

Radiolarians, 8, 251–257, **251**, **252**, **253**, **254**, **255**, **256**, **259**

Radó, Tibor, 159, **159**, 160, 168
Raphael, **82**
Rain drop, 240, **240**
Rational number, 50–51
Rayleigh, Lord, 215
Réaumur, R.A.F., 226
Reisch, Gregorious, **57**
Reflection
 at a curve, **37**, 73
 by curved mirrors, 75
 by an ellipse, 76, **76**
 by flat mirrors, 73, 198–199, 223
 law of, 73n, 73–75
 of light, **37**, **37**, 72–75, 77–81
 by a parabola, 77, **77**
 at a straight line, 73–75
Refraction of light, 123–125
 law of, 123, 125
Relativity theory, 297–300
Rest, law of, 19
Rhizophlegma, **255**
Riemann, Bernhard, 89, 159, 198n, 248, 298
Risner, Friedrich, **78**
Rococo period, 17
Roller, 130
Rømer, Ole, 72, **124**
Rome, 69, 77, 86
 aqueducts, 133–134
Rosetta Nebula, **295**
Rotating celestial bodies, 244–248
Rotating drop, 248–251, **249**
Roulade of a focus, 236–237, **236**
Royal Society, 274, 277, 278, 286
Rudolf II, Emperor, 61
Rudolphine tables, **62**

Saddle, **30**, 137
Saddle point, 172–175, 250
Saddle surface, 168, 203
St. Peter's cathedral in Rome, 132
St. Petersburg, 34
Sakkas, Ioannis, 81
Sambuca, 80
Samos, 48
Satellites, 247, 249, **249**

Saturn, 46, **58**
Saturnalis, **251**
Sayers, Dorothy L., 85
Scheiner, Christoph, 244
Schoen, Alan, 200, **200**, 201
 Schoen surface, **201**
Schumacher, 109
Schwarz, Hermann Amandus, 87, **87**, 159, 183, 185, 186, 191, 198, 198n, 200, **200**
 Schwarz chain, 198
 Schwarz periodic minimal surface, 200, 201, **200**, **201**
 Schwarz's problem, 87–91, **90**
 Schwarz's reflection principles (Rules 3 and 4), 199
 Schwarz's rule (Rule 1), 183, **183**, 186
 Schwarz's solutions to the Plateau problem, **159**
 Schwarz's triangle. *See* Altitude triangle
Screw surface
 motion, 180
 See also Helicoid
Self-gravitation, 219, 241–248
Semicircle, 64, **67**, 218
Set of coincidence, 208
Shakespeare, William, 9, 41
Shortest connection, 85–125, **101**, **114**, **115**, 116–119
 of several points, 92–109
 of two points in a plane, 46, 74–75, **74**, **75**, **110**
 of two points on a sphere, 46, **46**, 116
 of two points on a surface, 110–120
Shortest line, 86, 110, 113, 114, 115
Shortest path, 20, 74, 92, **97**, **101**, **110**
Sicily, 72, 77, 82
Siege of Syracuse, 77–83
Simply connected, 241
Simpson, Thomas, 98
Single-sphere pit, 250, **250**

Skeletons
 of diatoms, **4**
 of radiolarians, **7, 8**, 251–257
Smoke ring, 12
Snell, Willebrord, 125
Soap bubble, 3, 14, **14**, 100, 142, 146, 219–224, 228–234, 267
 within a tetrahedral frame, **8**
Soap Bubbles, Their Colours, and the Forces Which Mold Them (C.V. Boys), 232
Soap film(s), 3, 100, 110, 142, 143, 145–211, 228, 252
 held by threads, 205–211
 in an octahedron, **196**
 rules for, 183, 192, 225, 253
 as surfaces of minimal area, 146–160
 as surfaces of zero mean curvature, 160–168
 in a tetrahedron, **8**
Soap film solution of Steiner's problem, 197, **197**
Sobral, 300
Solar eclipse, 243, 300
Solar system, 247
Solstices, 284
Solution(s)
 of the brachystochrone problem, 121, 123–125
 of the horse-watering problem, 75
 of the Plateau problem, 158–159
 of Schwarz's triangle problem, **87**, 87–88
 of Steiner's problem, 92–109, **97, 101, 102**
 uniqueness, 100
Spacetime, **273**, 299, **298–299**
Speed of light, 72
Sphere, 1, 46, 79, 111, **111**, 113, 114, 115, 150, 151, **151**, 152, **153**, 231, 234, 252, 266, **266**
Spherical motion, 57
Spheroid
 oblate, **245**, 245–247
 prolate, 245

Spinel, 263
Spinning ball, 248
Spiral Galaxy, **246**
Spring, 139, **139**
Square root of two, 50, 51
Stability limit, 178, 232, 241, 246
Stable configurations, 146, 232, 241
Stable equilibrium states, 138
Stable minimal surface, 146–160, 167, 170, 172, 175
Stars, 244
Statics, 130, 132
 fundamental law of, 137
Stationary states of equilibrium, 138, 143
Stationary motion, 292
Statique expérimentale et théoretique des liquides (Plateau), 147
Staurocaryum, **255**
Staurolonchen, **255**
Steel ball on trampoline, **298–299**
Steepest ascent or descent. *See* Principle of steepest ascent
Steiner, Jacob, 92, **92**, 218
 Steiner's problems, **55**, 92–109, 193, 197, **197**, 210, 218, 223
 Steiner tree, 104, 106
Stenius, 186
Stevin, Simon, **126**, 128–130
Straight line, 46, 52, 53, 61, 73, 110, 236, 237
Straight rail, 53
Stuttgart, Institute for Light Weight Structures, 207, **209**
Sumerians, 43
Summit, 29, **29**
Sun, 46, 57, 58, 61, 77, **78**, 189, 243, 244, 274, 275, 276, 284, 300
 dark spots of, 244
Supporting surface, 182, 186, **187**, 191, 229
Surface(s), 112, 150–154
 annular, **176**, 177
 closed, 150
 complete, 118, 233, 234
 concave, **37**, 112, **117**

of constant mean curvature, 167, 231–233, 235–237, 252, **252**
convex, **37**, 117, **117**, 119
disc-type, 154, 158, 159
finite, 233
with holes, 151–153, **153**
of least area, 146–211
of minimal area, 146–211
nonorientable, 156–158
one-sided, 156–158
orientable, 156–158
of revolution, 113, **113**, 116, 177, **177**, 178, **178**, 252, **252**
two-sided, 156–158
with self-intersection, 199, 234, **235**
without self-intersection, 160, 200, 234, **235**
Surface area, 79
 reduced, 223
Surface tension, 142, 160, 226, 242, 248, 252
Symmetrization of a convex curve, 217
Syracuse, 77, 79, 80, 82
System of soap bubbles, 14, 228–230
System of soap films, 177, **177**, 191–197

Tallquist, 186
Tangent line, 70, 94, 163
Tangent plane, 117, **117**, 138
 horizontal, 29, 138
Tarentum, 52
Tartaglia, Nicolo, 130
Tautochrone, 122
Tensile stress, 204
Tetrahedron, 197
Thales, 42, 44, 45, 46, 218
Theatetus, 57
Theodicy, 22–23, 300–302
Theorema egregium, 115
Theon, 69n
Thompson, D'Arcy, 226, 227, 251, 252, 253
Threads, 142, 229, **231**
Thread experiments, 206–211

Three-lobed figure, 249, **249**
Tides, 280
Tightest packing, 225, **225**
Tin oil, 10
Toeplitz, Otto, 132
Tongue, 188, **188**
Topological type (of a surface), 150–160, 176, 181, 186, **186**
Topology, 175
Toricelli, Evangelista, 95, 135
 circle, 95, **95**, 96
Toricelli, Evangelista (continued)
 point, 96, 97, 98, 99, 102
 system, 102, 103
 tree, 103, 104, **105**, 106, **106**
Torsional moment, largest, 215
Torus, 152
Total curvature (of a curve), 169
Toulouse, 123
Trampoline model, **298**, 299, **299**
Transcendental numbers, 50
Trebizond, George of (Trapezant), **59**
Tree, 103
 phylogenetic, 106
 See also Toricelli tree and Steiner tree
Triangle(s)
 acute, 87
 altitude, **87**, 87–88
 inscribed, 87, **87**, 87n
 isosceles, **49**
 light, 87
 right, **43**, **49**, 50
Triangulation, 266, **266**
Tridiktyopus, **252**
Trochoid, 53, **53**
Tsetses, 80
Tyre, 64

Tycho de Brahe, 61, **62**

Ubaldi, Gaido, 135
Unduloid, 236, **236**
Universities
 Basel, 34, 119
 Berlin, 69n, 87
 Bologna, 59
 Cracow, 59
 Göttingen, 61
 Graz, 61
 Heidelberg, **131**
 Leipzig, 33
 Tübingen, 61, **63**, 156
Unstable configuration, 232
Unstable equilibrium state, 137
Unstable minimal surface, 160–168, 175, 178, 232

Van Schooten, Frans, 285
Variational principles, 39, 63, 262, 269, 273, 286
 in dynamics, 286–293
 See also Minimum principle
Varignon, Pierre, 135n
Vasco da Gama, 86
Vault, 132
Velocity, 20, 270, 273, **273**, 274, 288
 angular, 274
Venus, 46, 58
Vergil, 64, 65, 69
Vertex, 191
 of a cone, 53
 of a polyhedron, 174
Viviani, Vincenzo, **55**
Voltaire (François Marie Arouet), 23, **23**, 24, 26, **30**, 33, 33n, 38, 245
Volume, 222, 223
 of a cone, 79
 of a cylinder, 79
 of a halfball, 79, 223
 maximum, 222
Volvox, **2**
Von Karman, Theodor, 11
Vortex street, 11
VSLI layout (of a computer chip), 106

Wallis, John, 273
Water droplets, 140
Water jet, 232
Wedge, 130
Weierstrass, Karl, 69n, 89, **89**
Wente, Henry, **234**
Wetting liquid, 238, 240
Whiteside, Derek Thomas, 271, 272, 285
Whirlpool, 11
 at Narúto, **11**
Whirlpool galaxy in Canes Venatic, 13
Whole number. *See* Integer
Wimsey, Lord Peter, 85, 89
Wire frame, 146, **147**
Witelo, **78**
Wolff, Christian von, 22
Woolsthorpe, 285, 286
Worm of bubbles, **228**, **229**
Wren Christopher, 273, 276
Wulff, G., **264**, 265, 266, 267
Wulff construction, 265–267, **267**

Xenophanes, 1
Xiphostylus, **251**

Zenodor, 69n
Zero mean curvature, 160–168, 177, 180, 231

GPSR Compliance

The European Union's (EU) General Product Safety Regulation (GPSR) is a set of rules that requires consumer products to be safe and our obligations to ensure this.

If you have any concerns about our products, you can contact us on

ProductSafety@springernature.com

In case Publisher is established outside the EU, the EU authorized representative is:

Springer Nature Customer Service Center GmbH
Europaplatz 3
69115 Heidelberg, Germany

www.ingramcontent.com/pod-product-compliance
Lightning Source LLC
LaVergne TN
LVHW080311260326
834688LV00038B/1051